SUPERCONDUCTIVITY

THE NEW ALCHEMY

JOHN LANGONE

CB

CONTEMPORARY
BOOKS

CHICAGO · NEW YORK

621.3
L28a

Library of Congress Cataloging-in-Publication Data

Langone, John.
  Superconductivity : the new alchemy / John Langone.
      p.   cm.
    Includes index.
    ISBN 0-8092-4581-7
    1. Superconductors.   2. Superconductivity.   I. Title.
  TX7872.S8L36   1989
  621.3—dc20   89-31950
                                                    CIP

"Anatomy of a Magship," illustration on page 117 by Michael Rowe
© 1985 Discover Publications. Used by permission of the publisher.

bc

Published by Contemporary Books, Inc.
180 North Michigan Avenue, Chicago, Illinois 60601
Manufactured in the United States of America
Library of Congress Catalog Card Number: 89-31950
International Standard Book Number: 0-8092-4581-7

Published simultaneously in Canada by Beaverbooks, Ltd.
195 Allstate Parkway, Valleywood Business Park
Markham, Ontario L3R 4T8 Canada

To the memory of Atsuko Chiba, journalist and friend

Other Books by John Langone

CONTENTS

PROLOGUE

IT IS THE year 2000, and Electric Man of 300 K Ceramic Drive, Superconductivity City, is impressed. He has just watched the launch—on his superconducting, high-definition, satellite TV—of yet another U.S. manned rocket to Mars. It had flung itself cleanly up and out of its superconducting, electromagnetic launching tunnel in a liftoff that, though minus the flame, smoke, and deafening noise of the "old days" of solid-rocket boosters, was still a thrill to behold. The craft's payload on this trip includes a battery of highly sensitive, superconducting quantum interference devices designed for geophysical prospecting on the Red Planet, and several superconducting bolometers and quasiparticle mixers to detect and measure microwave and far-infrared radiation.

As the rocket ship, guided by superfast superconducting computers, each the size of a sugar cube, shoots out of the range of earth's gravity, Electric Man presses a button in the arm of his easy chair. The chair lifts a few inches off the floor and then, responding to the touch of another button, glides noiselessly on a cushion of air over the tiny superconducting magnets embedded in pencil-thin stick-on tracks that run between the TV set and a desk in the corner.

Electric Man picks up his appointment book and glances nervously at his superconducting wristwatch (a one-time electrical charge and a drop of liquid nitrogen every year or so is all it needs to keep it running forever). He has an annual doctor's appointment and a train to catch, and it's three hundred miles to the regional biomagnetic medical center. But it's a short run to the station, and Electric Man's superconducting car—its pollution-free engine driven by an almost unlimited supply of electricity stored in coils of wire that require no recharging—can do eighty miles per hour if pushed.

As Electric Man heads his car silently down the highway, he passes a gigantic, doughnut-shaped, grassy mound that hides the underground superconducting ceramic coil, which supplies his home and office with electric power. Chilled by a shroud of liquid nitrogen, the huge "battery"— the size of a football stadium—stores hundreds of megawatts of power, produced by superconducting generators hundreds of miles away, during nighttime periods of low customer demand. The energy has been sent economically and efficiently to the storage coils through underground, resistance-free ceramic cables. Because the coils are also superconductors and offer no resistance to the energy flow, the current, once it is pumped in, will circulate indefinitely without resistance until it is tapped. All of it will always be available for use.

Electric Man smiles when he thinks that only a few years ago, the power companies were losing an average of $200 million a year on wasted electricity because 15 percent of what they generated was lost battling the resistance in plain old copper wire.

As Electric Man pulls up to the station, the maglev train, a gleaming fiberglass vehicle that looks like a 747 minus the wings and jet engines, sits in a trough-shaped guideway that is studded with aluminum electric coils. The train has no wheels and no engines. Instead of rolling down the rails as its predecessors did, it literally flies, suspended a few inches above the coils by superconducting levitation magnets in its undercarriage—much larger versions of the ones that lift Electric Man's easy chair—and driven smoothly and noiselessly forward by propulsion magnets. Indeed, Electric Man will be at the medical center in less than an hour—a trip that would have taken him close to four hours before the likes of Amtrak's Flying Carpet came along.

When Electric Man arrives at the medical center, he takes off his shoes and steps into one of the pairs of magnetized, superconducting clogs lined up by the door. A nurse tells him to get on the blue track and ride it to the end of the corridor, step off onto the red track, and follow that to the mag lab, a metallic room that will shield him from outside magnetic fields. Electric Man floats effortlessly there, and a few minutes later he is lying on a low examining table, waiting for the doctors who will take his magnetoencephalogram and his magnetocardiogram.

Electric Man's neurologist enters the room, presses a button on the wall, and the examining table rises a foot off the floor and glides over to a bank of sensors that resemble hair driers. The table stops under one and gently settles back onto the floor. The sensor, lined with ultrasensitive superconducting detectors, now descends slowly to fit neatly around Electric Man's head. The doctor throws a switch, and in a few seconds the faint magnetic fields that emanate from Electric Man's brain cells are precisely mapped and displayed on a video screen. "I can give you a clean neurological bill of health just by eyeballing this reading," says the doctor. "The fields are uninterrupted. That means no epilepsy, no tumors, no emotional depression, no proneness to stroke. You have some minor hearing loss, a touch of short-term memory loss, and a little myopia, but those are normal for your age."

Electric Man's table rises again, and he is transported

out the door and into an adjoining room, where a cardiologist and an internist guide the table under another superconducting sensor, this one shaped to fit Electric Man's entire body from the neck down. Again, the magnetic fields that flow out of Electric Man's body, a billionth of the strength of the earth's, are monitored.

"Electrical action of the heart, normal," says the heart specialist. "No danger of a heart attack for you."

"Kidney, bladder, and liver function fine," says the internist, perusing the magnetic map. He smiles. "You must have eaten a can of something for lunch, because we're picking up some activity from the metal particles your can opener left, and I bet you had some acupuncture for that arthritis in your left hand not long ago. The needles those guys use are sometimes electrified, and the disturbance pattern is a residual effect. There's a slight break in the field from your duodenum that's indicative of the beginning of a small ulcer, but we can handle that easily. See you next year."

SUPERCONDUCTIVITY MILESTONES

1911 Dutch physicist Heike Kamerlingh Onnes discovers superconductivity in mercury at temperatures of 4 kelvins (4° K).

1913 Kamerlingh Onnes is awarded the Nobel Prize in Physics for his research on the properties of matter at low temperatures.

1933 W. Meissner and R. Ochsenfeld discover the Meissner Effect.

1937 German physicist Fritz London speculates that supercurrents may exist in nonmetallic systems.

1941 Scientists report superconductivity in niobium nitride at 16° K.

1953 Vanadium-3 silicon found to superconduct at 17.5° K.

1962 Westinghouse scientists develop the first commercial niobium-titanium superconducting wire.

1972 John Bardeen, Leon Cooper, and John Schrieffer win the Nobel Prize in Physics for the first successful theory of how superconductivity works.

1986 IBM researchers Alex Müller and Georg Bednorz make a ceramic compound of lanthanum, barium, copper, and oxygen that superconducts at 35° K.

1987 Paul Ching-Wu Chu of the University of Houston and Maw-Kuen Wu of the University of Alabama substitute yttrium for lanthanum and make a ceramic that superconducts at 98° K, bringing superconductivity into the liquid nitrogen range.

1988 Allen Hermann of the University of Arkansas makes a superconducting ceramic containing calcium and thallium that superconducts at 120° K. Soon after, IBM and AT&T Bell Labs scientists produce a ceramic that superconducts at 125° K.

1

GOING FOR THE COLD

ONE FATEFUL DAY in 1980, as the people down at the Institute of Electrical and Electronic Engineers like to tell the story, Rustum Roy, a physical chemist at Penn State, became disenchanted with his experiments in superconductivity, which is the ability of some substances, when cooled to very low temperatures, to conduct electricity without resistance and without loss. He had been experimenting for five years with ceramics—notably with a barium-lead-bismuth oxide mixture—but despite the long hours and the hard work, he could not get his concoction to superconduct at temperatures any higher than a few degrees above what one might encounter in outer space.

"I'm not wasting any more time on these damn superconductors," Roy told Bernd Matthias, a professor at the University of California at San Diego, the discoverer of a number of superconducting materials and the man who had steered Roy toward the barium mixture in the first place. Matthias was persistent. "Don't give up, Rusty," he said. "This is a completely new type of superconductivity."

That night, after a game of tennis, Matthias went to bed early and died in his sleep.

Matthias, it now turns out, was an incredible prophet. The very recent discovery of a family of ceramic superconductors similar in crystal structure to the compound with which Roy had been experimenting has galvanized the interests of researchers throughout the world. For not only do these ceramics superconduct in a strange new way, but they do so at temperatures higher than ever before dreamed possible. Some scientists even predict that yet-to-be-discovered ceramics will superconduct at temperatures at or above room temperature, a development that would represent a breakthrough of such enormous proportions that it would drastically change the very way we use electricity.

Indeed, the number of potential applications is mind-boggling: highly efficient power generators; superpowerful magnets; computers that process data in a flash; supersensitive electronic devices for geophysical exploration and military surveillance; economic energy-storage units; memory devices such as centimeter-long video tapes with superconducting memory loops; high-definition satellite television; highly accurate medical diagnostic equipment; smaller electric motors for ship propulsion; magnetically levitated trains; ore-cleansing magnets; more efficient particle accelerators; fusion reactors that would generate cheap, clean power; and even electromagnetic launch vehicles and magnetic tunnels that could accelerate spacecraft to escape velocity.

And those are just the potential applications that are currently being extrapolated from technology already under investigation for the "ordinary" superconductors, the low-temperature ones that have been around for years. There are sure to be others that no one has thought about yet.

Just what is it all about, this superconductivity, this electrical example of perpetual motion that has stimulated such fierce competition among the leading industrialized

countries and grabbed headlines so often? First, its phenomenal power should be put into perspective: for all of its
enormous potential, it will be some time before superconductivity can be harnessed and made to do all that its
acolytes say it can do. For another, superconductivity—
despite its current buzzword status—is not really all that
new. It was discovered in 1911 by the Dutch physicist
Heike Kamerlingh Onnes, and superconductors based on
his pioneering research have been commercially available—albeit in but a few applications—for years.

Thus, when one speaks of superconductivity these days,
it is more accurate to refer to it as the *new* superconductivity. But while it is not stretching the truth very much to say
that Kamerlingh Onnes was proof of Andrew Carnegie's
dictum that "the first man gets the oyster, the second gets
the shell," it is also true that despite the Dutchman's
ground-breaking discovery, superconductivity languished
as a sort of scientific curiosity for some seventy-five years.
This is not to say that Kamerlingh Onnes's accomplish-
ment was trivial. If it had not been for his basic, dogged
research, our understanding today of the strange phenomenon—not to mention the discovery of a host of new materials that make superconductivity happen at unheard-of
temperatures—would have been longer in coming. No
problem can be solved, be it scientific, social, or economic,
without first doing the groundwork.

Kamerlingh Onnes's research languished, but fortunately it was not entirely forgotten. A relative handful of
scientists had the foresight as well as the thickness of skin
to ward off their colleagues' aspersions and haul superconductivity out of the dead-end street it had gotten into,
gaining for themselves the last laugh. For this small group
of researchers, superconductivity was not just a fling with
the arcane, an attempt to learn more and more about
matters of lesser and lesser importance. Nor was it an
opportunity to perform stunts of laboratory fireworks that
would, like the waxed wings that bore Icarus to the sun,
impress the observer mightily for a flash before losing their

bright promise. Instead, for these researchers Icarus was, in the words of the English astronomer and physicist Sir Arthur Stanley Eddington, "the man who brought to light but a constructional defect in the flying machines of his day."

"Icarus," said Eddington, "will strain his theories to the breaking point till the weak points gape. For the mere adventure? Perhaps partly; that is human nature. But if he is destined not yet to reach the sun and solve finally the riddle of its constitution, we may hope at least to learn from his journey some hints to build a better machine."

Kamerlingh Onnes had also, quite literally, reached for the sun before he became interested in the resistance of metals to electrical current. Specifically, he'd focused his attention on the inert gas helium, whose name is derived from the Greek word for sun, *helios*. It was his research with helium that opened the door to the discovery of superconductivity. Colorless and odorless, the lightest of all gases other than hydrogen, helium had been discovered during an eclipse in 1868 by the French astronomer Pierre Janssen, who noticed some strange lines, apparently unrelated to any then-known substance on earth, in the emitted light of the hot incandescent gases in the sun's atmosphere. A quarter century later, Sir William Ramsay, the Scottish chemist who was best known for isolating elemental gases from the atmosphere, became the first to isolate helium successfully from terrestrial sources. At about the same time, radioactivity was discovered, and not long afterward, Sir Ernest Rutherford, among the first and foremost of nuclear physicists, placed the gas among the decay products of a radioactive element's disintegration.

But there was more to be done with helium. By 1907, about the same time that Rutherford was connecting it to radioactive elements, all known gases except helium had been transformed into liquid by scientists eager to learn more about their behavior, which was, along with their structure, far simpler than that of elements occurring naturally as liquids and solids. (Most gases are biatomic; that is, they have but two atoms in their molecules. Helium has one.) Early on, the English chemists Sir Humphry Davy and Michael Faraday had managed to liquefy several familiar gases—among them hydrogen sulfide, chlorine, and carbon dioxide—by placing mixtures of chemicals that generated the gases in one end of a sealed glass tube shaped like an inverted V, and the other end into a freezing mixture. By heating the chemicals, they produced gas, which was caused to liquefy by the combination of cold and pressure.

Researchers began to suspect that there was no such thing as a "permanent" gas, that it was possible to liquefy all of them, including helium, by putting them under pressure and cooling them below a certain temperature peculiar to each, called the critical temperature. This is the highest temperature at which a gas can exist in a liquid state and above which it would boil away, that is, evaporate back into its natural gaseous form when exposed to an environment at room temperature. Moreover, they were to discover that once gases were liquefied, they became supercooling agents with vast commercial and scientific application.

Yet, for a number of reasons, helium had defied liquefaction. For one thing, while it was a major constituent of the sun and other stars, helium was difficult to find on earth. Even more important was the problem of its critical temperature, which seemed to be far colder than any temperature that had been attained and employed thus far. Chlorine, for example, was readily liquefied at a pressure of 100 pounds per square inch at a balmy 68° F. But even those gases with incredibly low critical temperatures had succumbed: oxygen, liquified at −297.4° F; nitrogen, −320° F; neon, −410° F; and hydrogen, −422° F.

Hydrogen, though the simplest and lightest element, had been especially nettlesome to the cryogenics scientists (cryogenics, from the Greek *kryos*, for frosty, is the science of producing and maintaining very low temperatures). Researchers knew that liquefying, or solidifying, any gas was now a matter of achieving lower temperatures, but how to reach some of the required subarctic readings presented some problems. Eventually, they devised ingenious "cascading" techniques that allowed them to subject the more stubborn gases to the chilling effects of previously supercooled gases, in effect cooling the more resistant gases by steps. They might, for example, use liquefied sulfur dioxide, a popular refrigerant, bleach, and disinfectant, to liquefy carbon dioxide (which, when frozen at $-109.3°$ F, forms our familiar dry ice), then use the liquid carbon dioxide to knock down the temperature of a resistant gas. Researchers also knew that gases could be cooled merely by letting them expand in a supercooled environment tightly insulated against any trace of outside heat. With that knowledge, and armed with supercold liquid nitrogen as the refrigerant, they were able to turn their attention to hydrogen.

In 1900, the Scottish chemist Sir James Dewar, co-inventor of cordite, a smokeless gunpowder, as well as the silver-coated glass vacuum flask that bears his name but is popularly known as a thermos, successfully cooled hydrogen to a frigid liquid by putting it into a sealed container surrounded by fluid nitrogen, a process which allowed the hydrogen to expand and, thus, cool even more. In doing so, Dewar obtained the lowest temperature reached up to that time. He managed a reading just a few degrees above $-460°$ F, that coldest of the cold, absolute zero, the theoretical temperature at which the atoms and molecules of a substance lose virtually all of their frantic, heat-dependent energy and at which all resistance stops short. Heat, at that temperature, is completely gone. (By way of reference, the coldest temperature ever recorded on earth was $-126.9°$ F, at the Soviet Union's Vostok research station on Antarctica.)

To better express the values of this unbelievably frigid world, scientists use what is called an absolute temperature scale that begins with absolute zero as its zero point; this is the Kelvin scale, named for William Thomson Kel-

COMPARATIVE TEMPERATURE SCALES

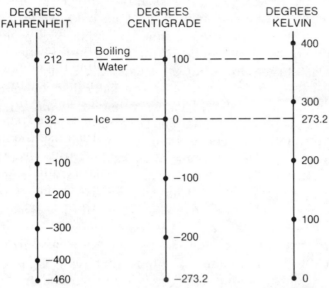

On this temperature chart, the zero reading on the Kelvin scale (right) represents absolute zero, the point at which heat is completely gone. Scientists studying superconductivity use the Kelvin scale to better express the values of the frigid world in which metals and ceramics superconduct.

vin, the Irish mathematician who devised it in 1848. (On the Kelvin scale, absolute zero is the equivalent of –460° F. Water freezes at 273° K, 32° F. Water boils, on the Kelvin scale, at 373°, which is comparable to 212° F.) Physicists believe today that it is impossible experimentally to reach absolute zero, but they've gotten pretty close, down to around 0.00001° K, by subjecting the nuclei of certain metal alloys to a magnetic field at extremely low temperatures; when the magnetic field is removed, the nuclei become demagnetized, and their temperature falls to almost absolute zero. (Even the inordinately cold regions of outer space register only 3° K.)

But to return to Kamerlingh Onnes and helium. A man of extremely delicate health and an enormous capacity for work—two seemingly incompatible qualities that in conjunction often create greatness—Kamerlingh Onnes was appointed professor of physics at the University of Leiden at the age of twenty-nine, a post he held for forty-two years. It was a transitional time for physics; its practitioners were coming around to the view that matter had a corpuscular nature—that is, consisted of small, discrete particles—and was not, as many scientists believed, a continuum, some sort of vague, coherent whole. Another Dutchman, physicist Johannes Diderik van der Waals, had been promoting the corpuscular theory, arguing, in the case of gases, that these amorphous will-o'-the-wisps were circulating molecules which exerted binding forces on each other and that all gases behaved in exactly the same way.

Kamerlingh Onnes, a devotee of quantitative research in physics—quantitative analysis being the measurement of the proportion of known components in a mixture—had been studying the conformity in the behavior of gases and had set out to obtain some precise measurements of gaseous states. To better see how gases actually behave, however, it was necessary to slow down their molecules to near immobility, and the way to do that was to work at very low temperatures. Kamerlingh Onnes built himself some cascading apparatuses, first liquefying methyl chloride and ethylene to liquefy large quantities of oxygen and air (air liquefies at $-317.9°$ F), and then using that to liquefy hydrogen. Finally, on July 10, 1908, he attacked the last holdout among the so-called noble gases, helium, using hydrogen to cool it under pressure. This allowed the helium to expand and thus to cool down even more. The painstaking process resulted in helium changing from a gas to a liquid at $4.2°$ K ($-452°$ F), only a few degrees above absolute zero.

Kamerlingh Onnes had only a thimbleful of helium liquid, but it was to be the grand overture to his explorations in a vast new temperature region, a place of intense cold where the physical properties of many substances changed remarkably. Even at the temperature of liquid air, one could perform strange tricks. One can, for example, mold the silvery white, normally liquid metal mercury into a hammer and drive nails with it, or bounce a superchilled ball of India rubber against a wall and shatter it like glass. Kamerlingh Onnes was, however, on a much higher flight than performing tricks in a physics lab. He had developed an interest in the electrical resistance of metals.

Scientists were beginning to suspect that when the temperature around conducting wires was lowered, something happened to electrical charges coursing along inside—something far different from what occurred when current flowed at room temperature, 68° F. Kelvin believed that electrons traveling through a conductor would come to a complete halt as the temperature got close to absolute zero. But others were not so sure. They felt that a cold wire's resistance—its inability to carry current—would dissipate. This suggested that there would be a steady decrease in resistance, allowing for better conduction of electricity, as the temperature went lower and lower and as the normal vibrations of the atoms in the wire slowed down. At some low point, the theorists said, there would be a leveling off as resistance reached some ill-defined minimum value, and current would flow along unimpeded.

To be sure that the metal wires with which he experimented would give him the best possible results, Kamerlingh Onnes cleansed them of impurities before subjecting them to low temperatures. He tried palladium, a white soft metal in the platinum group, and he noticed that the less foreign material the wires contained, the lower the resistance got before it leveled off.

In order to lessen the influence of impurities even further and perhaps get better conductivity at a still-lower leveling-off point, he hit upon mercury. Because it was a liquid at room temperature, it was easier to clean up, and the slippery liquid could be molded readily into wire, in which resistance could then be measured accurately.

Kamerlingh Onnes made his mercury wire and saw a gradual decrease in resistance as he steadily lowered the temperature and played around with current. But he wasn't quite ready for what happened when he supercooled the wire with liquid helium and sent current through it. There was no leveling off—let alone any stopping of electrons as Kelvin had suggested. Instead, there was a total disappearance of resistance. Current was flowing through the mercury wire, and nothing was stopping it.

Kamerlingh Onnes called his discovery a new state of matter, and named it supraconductivity.

2
THE NATURE OF CURRENTS

MOST OF US probably know that copper, wound in coils or strung in wires and cables, carries electricity, and that sometimes aluminum does the trick. But chances are our knowledge of normal conductivity ends there. As familiar as all the uses of electricity are, many of us still haven't the foggiest idea about what actually goes on after we plug in the cord of a lamp or an electric drill and flick a switch, or twist the ignition switch in a car, or walk across a rug in a pair of crepe-soled shoes and feel and hear a *zzzzzt* when we reach for a fork. Many of us would also be hard-pressed to explain the interrelationship of electricity and magnetism, or define volts, watts, ohms, and amps. Even the household word *current* might elicit only a few brief words, most of them analogous to water flowing through a pipe.

So it's no wonder that for all the media hype surrounding it, superconductivity—the very name conjures up something that belongs in a "Star Trek" film—should be a mysterious occurrence understood only by brainy scientists working in laboratories shot through with crackling lightning charges.

But for the moment, let's forget the "super" part of conductivity and just concentrate on the rest of it. Conductivity refers, of course, to the ability of a substance to carry electricity, but for now we'll talk only of those that conduct under ordinary, not supercooled, circumstances. Some of the garden-variety conducting substances do it well, like copper and aluminum, and also silver and gold. Electrolytic solutions and ionized gases are also conductors. (Electrolytic solutions contain chemical compounds that have become ionized—that is, their molecules become electrically charged. Ions may also be formed in gases by radioactivity, by the sun's ultraviolet rays, or by electrical discharge.) Some substances conduct only halfheartedly—for example, germanium, a brittle, metalloid element; and silicon, nonmetallic and the most abundant element found in the earth's crust apart from oxygen—and these are called semiconductors. Some substances are unable to carry electricity at all—mica, rubber, porcelain, and glass among them—and these are the insulators.

To understand conductors, one has to understand the nature of what flows through them—electricity. Benjamin Franklin regarded electricity as a sort of fluid existing in all matter and whose myriad effects could be explained by too much or too little of it. Awareness of some of those effects goes back a long way, at least as far as 321 B.C., when the Greek philosopher Theophrastus mentioned the peculiar ability of amber (a yellowish fossil resin): when rubbed vigorously, it attracts straw, dry leaves, and bits of lint. The Greek word for amber is *elektron*, and it is from this that we get our term *electricity*.

Theophrastus was talking about what we know now as one of the two forms of electricity, static electricity, in which charged particles are transferred from one substance to another, where they regroup and remain in a sort of resting state. The other form, current electricity, occurs when the particles are in motion, as when they move through a wire.

The charged particles to which we are referring are protons and electrons, fundamental bits of matter, and it is the various related phenomena caused by their relative position and relative movement that we call electricity. Protons are positively charged and are found in the nucleus of the atom, its central part, which comprises nearly all of the atomic mass. Electrons are negatively charged, and various numbers of them orbit around the nucleus in shells, at dizzying speed, making one of billions of trips in a millionth of a second. Each electron carries a single unit of negative electrical charge, each proton one of positive charge. Generally, atoms have equal numbers of protons and electrons, making the atoms electrically neutral. The number of protons in each atom represents its atomic number, and defines the chemical element. Helium, for example, is the lightest and smallest atom, with one proton and one electron, and bears the atomic number 1; copper's atomic number is 29, that is, 29 protons, 29 electrons; uranium's is 92.

But while atoms are usually electrically neutral, this delicate balance can be upset if they gain or lose electrons; when this happens, the atoms become electrically charged and are called ions. The charge can be positive or negative, depending on whether there is an excess of electrons or a deficit. In the simple case of amber, it becomes negatively charged when rubbed with fur or wool, and will attract small objects. It picks up its charge because it now has gained an excess of electrons, electrons which, to begin with, were loosely held in the fur, which is now positively charged. The amber will now attract the fur because of the fundamental, familiar electrical law: unlike charges attract. When a glass rod is rubbed with silk, it too gives up some of its easily detached electrons, becomes positively charged, and able to attract objects. But when two glass rods are rubbed with silk, which transfers their free electrons to the fabric, they repel one another because both have positive charges. Another fundamental electrical law is at work: like charges repel.

In these examples of static electricity, all that is going on, as we noted earlier, is a simple reshuffling of electrons as the charges move from one substance to another, leaving one with a negative charge, the other with a positive one. No real flow of electricity, actually, just a deposit of sorts, from here to there. But despite its apparently uninteresting role as a mere collector of positive or negative charges, there is an important aspect of electrostatics to be considered: the electricity associated with it can be put in motion—made to follow a current. (We will focus only on the negative electrical charge, since it is the one that turns on our lights and television sets and drives our motors, while the positive charge manifests itself rarely, under special circumstances.)

Perhaps the first person to construct an electrostatic generator, and a dribble of electrical current along with it, was Otto von Guericke, a seventeenth-century German diplomat and engineer, and mayor of Magdeburg. He built a sphere of sulphur and, by rotating it and rubbing it at the same time with his hand, found that it not only attracted shreds of paper and cloth, but that it gave off sparks. Moreover, when he attached a linen thread to the sphere, it acquired the same ability to attract as the sphere. Although from all accounts von Guericke didn't recognize that what he was witnessing was due to static electricity—many of the early researchers regarded the phenomenon as a trivial one—it was, nonetheless, the beginning of a serious, experimental look at electricity.

Like von Guericke, some of the electrical pioneers lucked into the annals of science while entirely misconstruing the results of their laboratory dabbling. One of these was Luigi Galvani, a professor of anatomy at the University of Bologna who doubled as a physicist. It is said that his wife, daughter of a distinguished member of the medical faculty at Bologna, noticed some peculiar muscular convulsions in a skinned frog when a nerve in the creature's leg accidentally touched a metal scalpel that lay on a table and which had become charged by contact with a nearby static-electricity machine. She called her husband, who, methodically repeating the process, got the frog's legs to twitch when he touched them with the charged scalpel. Elated, Galvani stuck wires of different metals into the frog's leg muscles, turned on his machine, and *voilá*, charged electrons hopped through the air, causing other electrons in the wires to move and stimulate the nerves in the animal's limbs. Later, he tied the legs of a freshly killed frog with copper wire and hung them over an iron balcony and waited for a thunderstorm to see whether lightning would produce the same effect. Whenever the flesh touched the iron, the legs twitched—even without lightning.

Galvani didn't know, of course, that a flow of electrons was behind the contractions. He concluded that the source of electricity lay in the nerve and that the wires, though essential to the experiments, were merely conductors. He was correct in associating the result of his experiments with electricity, but wrong in assuming it proved the existence of some sort of animal electricity hoarded in the frog's muscles. Galvani is generally regarded as the first scientist to encounter the effects of an electrical current, and his name, furthermore, is inextricably linked to galvanization,

VOLTA'S BATTERY

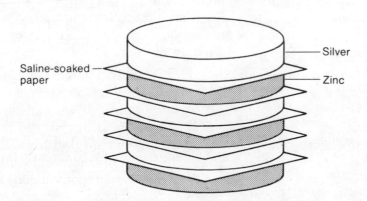

Volta's Battery, a simple pile of silver and zinc discs separated by brine-soaked paper, was the first device to generate a continuous electric current.

the process by which iron or steel are coated with zinc to protect them from corrosion.

Shortly after, Alessandro Giuseppe Antonio Anastasio Volta, an Italian physicist, invented the battery (the voltaic pile), and generated the first man-made, continuous electrical current. His device was crude, a simple cakelike stack of saline-soaked paper sandwiched between silver and zinc disks. But with this shabby-looking heap, he managed to prove that it was the action of moisture and these dissimilar metals, these inanimate materials, that generated the flow of current. Galvani's frog legs had only been a source of moisture, and the current the Bolognese had observed was not, therefore, any special property of animal matter.

In some electrical conductors, charges are free to move about—electrons in metals, or ions in salt solutions—like the molecules of a gas. A simple caricature has these looser electrons circling in the outer, or valence, shells of atoms, where the hold of the nuclei is weaker—much as a more distant planet is less strongly bound to the sun by gravity than the innermost planets. The solar system analogy, while good for a quick read on this topic of conductivity, is not, however, exactly appropriate as we get more deeply into the subject. Put another way, electrons occupy energy states (levels and bands) in the atomic network of solids that are peculiar to each material. Each level has a certain energy quota, and for an electron to move within that level, it must have the right amount of energy to qualify it for admission. Energy levels that run close to one another are called bands, and in most solid materials they are so close they almost blend.

Electrons can move between the various levels—and between different materials so long as their levels and bands overlap to create a "tunnel"—but there are two requirements: they need a boost of energy to reach a higher energy band, and a vacant slot in the band they enter. The outer shells of metallic conductors have several of these vacancies, along with electrons that are loosely held, and it is these slots that those free electrons head for when voltage—the energy, the electromotive force derived from an

ENERGY STATES

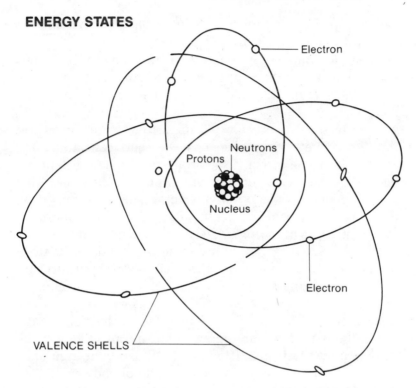

The conduction of an electric current involves electrons moving from one empty "slot" in an orbit around a nucleus to another empty slot. Here, electrons are shown orbiting a nucleus, and just to the left of the nucleus, a "free" electron is heading for an empty "slot."

electrical source, such as a generator or a battery—is applied. The conduction of an electrical current, then, is simply electrons moving from one empty slot to another in the atoms' outer shells, their orbital states.

It is the number of free electrons in a substance that determines how good a conductor it is. Current will flow much more easily through a substance, a conductor, that has at least one free electron per atom. In semiconductors—neither good conductors nor good insulators—the energy levels are filled at low temperatures, and no electrons are free to move about; but as temperature increases, some electricity will flow in certain directions as a few of the electrons pick up enough energy to enable them to make the jump into an empty conduction band. Insulators are constructed of atoms whose energy bands are completely full, with electrons tightly bound; when voltage is applied, electrons have no empty slots to move to, and current does not flow.

As current flows along—much like water trickling through gravel as the electrons leap between stationary atoms—it can be direct (DC), or alternating (AC), depending on the source. Direct current is produced by batteries and DC generators, and it always moves in the same direction; it is used to run the electrical systems of automobiles, and some types of motors. Alternating current, on the other hand, reverses direction many times a second and is used almost universally for domestic and industrial purposes in, among other things, radios, TVs, and other electronic devices.

AC has a distinct advantage over DC in that its voltage can be boosted far more easily than that of DC. Because electric power is transmitted most efficiently at high voltages, AC is the current of choice for the big power companies. AC voltage can also be decreased more easily than DC can. High voltages are unsafe for home and office consumption and for most factories, but with the intervention of a device called a transformer, electricity with low current and high voltage can be converted into electricity with high current and low voltage (or the other way around when need be) with practically no loss of energy.

Some definitions of a few of the numerous electrical measurement terms might be helpful here. Let's begin with the volt, which is generally defined as the unit of electrical potential. When voltage from a source like a battery is applied to a conductor, free electrons flow to form a current. The battery triggers an electromotive force (emf) by transforming its stored chemical energy into potential electrical energy; thus, one can say that a battery has a difference in potential energy between its ends that forces electrons to move along in a conductor. It is the emf, the pressure that causes the current to flow, that is measured by the volt. Most electric power for consumers in the United States and Canada comes into homes and businesses at 110 to 220 volts, and all general-purpose circuits are 120 volts. Typical power-plant generators produce electrical power at up to 22,000 volts, boosting it with transformers to as high as 765,000 volts for transmission.

Amperage measures the amount of electrical current, and its unit is the ampere, or amp. Electrical engineers have calculated that when an emf of 1 volt is hooked up to a conductor, it shoves 6,250,000,000,000,000,000 electrons past a given point in 1 second; every time this horde of electrons passes that point, 1 amp of current is recorded. An electron pack may move along fairly slowly in some currents because the amperage is high but voltage low; other currents move more swiftly, with high voltages and low amperage. When both are high, the result can be fearsome: a bolt of lightning can log in at 100 million volts and over 150,000 amps.

Watts are units of power, and wattage is the amount of power that is dissipated when energy is used. An electric clock might require 2 watts; an electric blanket, 200; a toaster, 1,000; an electric clothes dryer, 4,500; and an electric water heater, 9,000. One kilowatt is the amount of electricity needed to light a 100-watt bulb for 10 hours; an average home uses between 600 and 750 kilowatts of electricity a month.

The unit used to measure electrical resistance, that obstacle in the way of current flow, is the ohm. It represents resistance itself, along with its accompanying loss of energy, which are the barriers that superconductivity overcomes.

Whenever an electric current flows, even in the best of the ordinary conductors, it encounters some resistance, which changes the electrical energy into heat. Feel the warmth of an electric drill's housing after continuous use and you have an idea of heat loss due to resistance.

Such depletion, minuscule by a consumer's standards

and hardly noticed at all, becomes more significant when one considers the losses suffered by utilities during the generation, transmission, and distribution of electric power: a typical AC generator is 98.5 percent efficient in producing electricity, but only 95 percent of this power reaches consumers. If operating efficiency were improved to 100 percent—say with a superconducting wire—the change might appear small, but over the long haul it would be quite significant. New high-temperature superconductors could reduce the costs of electrical generators by some 60 percent, cut operating costs of large electrical motors by as much as 25 percent, and give consumers the equivalent of some 15 percent additional generating capacity by enabling the utilities to use their existing facilities far more efficiently.

Sometimes, the resistance a current encounters while passing through a conductor works to our advantage, as when the heat it produces is put to work. Plug in a toaster, for example, and even though the same current moves through the wires of the cord and the special alloy wires inside the appliance, only the latter glow red-hot. Because the resistance in the cord is small, the current creates only a small voltage, or pressure. In the toaster wiring, however, the resistance is made deliberately large, and because the voltage and the heat are also large, the temperature increases in the wires until they give off the heat to do their job. The electric light bulb is, of course, another familiar example: resistance in the filament causes it to glow. These are all examples of Ohm's Law, set forth by the German physicist Georg Simon Ohm, which states that the resistance within a conductor controls the amount of heat that develops.

But what, actually, is going on inside a conductor when resistance is encountered? Free electrons and the energy bands they travel in and between are the key. In metals like copper and aluminum, electricity is conducted by electrons that are jarred loose from their outer orbits and, moving as individuals, leap from one atom to another. These atoms form a vibrating lattice within the metal conductor—as difficult as it is to envisage, a hunk of metal, like all seemingly immobile objects, is actually a moving mass of atoms and molecules—and the warmer the lattice, the more it shakes. (If the lattice gets really hot, it shakes so much that molecular chunks break loose, the forces that have held it all together start to lose their grip, and the whole system quits being a solid and melts.)

As the electrons begin moving through this maze, they occasionally collide with tiny impurities—traces of iron, say—or with imperfections in the lattice. The lattice's vibrations make it even more difficult for the electrons to maneuver through. When the electrons bump into these obstacles—their trip is by no means smooth, but more akin to a careering pinball—they fly off in all directions and lose energy in the form of heat in the process. Such resistance to motion is also more familiarly known as friction. It is all somewhat like a lead marathon runner negotiating his way through a crowd of spectators who are blocking his path; he manages to get through, but each time he runs into someone he loses momentum—and raises a sweat to boot. That's resistance.

It is a vastly different scene inside a superconductor. The jungle gym, the lattice, is still there, and so are the impurities and imperfections. But the procession of the superconducting electrons through the obstacle course is different: they pass unobstructed through the complex architecture of the conductor. Because they bump into nothing and create no friction, the superconducting electrons have an enormous advantage over their kin traveling through ordinary conductors: they can transmit electricity indefinitely, with no appreciable decay in the current and no loss of energy. They zip smoothly through their supercooled wires like slalom racers, taking advantage of every opening, their sure course guaranteed so long as the trail is colder than ice.

What is it that endows the electrons in a superconductor with such uncanny skill when their equally able counterparts in ordinary conductors keep running into every obstacle that appears? One factor is the state of the lattice.

We've said that the warmer it is, the more it shakes, and the easier it is for electrons to smash into its struts. The converse of that is that the colder it is, the less it shakes, thus making it easier for electrons to get through.

But it's a bit more complicated than that.

The modern theory of superconductivity was advanced in 1957 by three American physicists—John Bardeen, who had won the Nobel Prize the year before for inventing the transistor; Leon N. Cooper, an expert in quantum theory (a theory developed in the 1920s to account for certain phenomena that could not be explained by classical physics); and John R. Schrieffer, who had specialized in electrical engineering before he switched to physics.

Long before that, though, the scientists who were carrying on Kamerlingh Onnes's work at Leiden were looking at possible mechanisms for superconductivity. Kamerlingh Onnes had been awarded the Nobel Prize in physics in 1913 for his research on the properties of matter at low temperatures, but although he ruled unchallenged over the field of cryogenics until his retirement in 1923, the theory behind the phenomenon he discovered remained a mystery to him. The scientists who had been continuing his work at Leiden were equally baffled, and their best guess was that this strange electrical movement had something to do with the ordering of the moving electrons in the conductor into some kind of condensed state.

An equally broad view was suggested in the 1930s by the physicist Fritz London, who came to England as a refugee from Nazi Germany. London, who with his brother Heinz had devised some equations that are still widely used to describe the electromagnetic properties of superconductors, speculated that electrons joined up in some way and moved as a team, like horses harnessed to a wagon, instead of singly, as a racehorse might travel down a track. The problem with this notion, however, was that electrons carry negative charges, and since like charges repel, it seemed that that repulsion would prevent the electrons from forming their team.

In the early 1950s, the German-born English physicist

Herbert Frohlich theorized a way out of the dilemma: free electrons interacting with lattice vibrations were involved in superconductivity. The idea was that the electrons created a slight distortion in the lattice, pulling it together, as they flowed through the conducting wire; this occurred because of the attraction between the electrons' negative charges and the positively charged ions in the lattice. That puckering, Frohlich suggested, acted as a positively charged wake, like the wake of a fast-moving ship, and other electrons following along behind were drawn into it and were, in effect, sucked along in a unified current. This relationship, then, hinged on the transfer of energy from an electron to the lattice, and from the lattice to another electron, and this interaction of energies supplied the momentum for a supercurrent.

Electrons, however, don't have a common velocity; in fact, laws of physics say that in a current, they cannot. How then could one expect that electrons would run in such harmony? Bardeen, who had independently visualized an electron-lattice interaction, sought to answer this and eventually realized that another ingredient was necessary to explain superconductivity. That ingredient, he suggested, was what is known as a condensation in velocity.

In 1956, after eluding explanation for nearly half a century, superconductivity was as open as a specimen in an anatomy laboratory, and the verdict was that it was another state of matter. In a sense, superconducting metals, which work their internal magic under the influence of intense cold and pressure, are the electrical world's counterpart to water, which changes its function and form when it becomes a solid at a mere 32° F.

The first forward step was taken by Leon Cooper, the slightly built graduate of the Bronx High School of Science in New York, a newly minted Ph.D. from Columbia, and the man Bardeen called the "quantum mechanic from the East." Cooper had joined with Bardeen and Schrieffer at the University of Illinois to tackle the mystery of superconductivity. (Schrieffer, from the Massachusetts Institute of

Technology, had been in Illinois before Cooper and had already been working on a problem dealing with conduction in semiconductors.) Because space was at a premium, Bardeen and Cooper shared an office. Schrieffer found a highly prized place with theorists, all graduate students, in what they jokingly referred to as the "Institute for Retarded Study" in a neighboring building. It was to be a most fruitful relationship, one that would win the three scientists the Nobel Prize for physics in 1972 for their theory of superconductivity—dubbed the BCS theory, after the initials of their last names.

Cooper and his team members knew that if electrons that ordinarily repelled one another did, indeed, come together in a superconductor, then some force that could overcome that repulsion and enable the electrons to coordinate their respective velocities had to be present. Cooper eventually hit on the idea that the lattice's vibrations were not directly responsible for unifying the whole current. Instead, they forced the electrons to pair up and move as a more formidable unit, a team that could get past all the obstacles that caused resistance in a conductor. (These electrons running in tandem are now known officially as Cooper pairs.) And what kept these particles that ordinarily would shun each other together? The answer to that was found to be phonons, packets of sound waves emitted by the lattice as it vibrated; phonons are units of motion, somewhat analogous to photons, the particles associated with light waves. Although this subatomic noise cannot be heard in the usual sense—if it could be heard, one might compare it to the tone that follows a struck bell—its role as a mediator is indispensable.

The whole process, according to the theory, goes like this: As one negatively charged electron passes by the positively charged struts in the lattice, the lattice distorts, sends out its phonons, and forms a sort of trough of positive charges around the electron. Before that electron gets by and before the lattice has a chance to spring back to its normal position, a second electron is drawn, or rolls into, the trough. What follows is one of those bizarre phenomena of quantum mechanics whereby the two electrons, which should repel one another, link up—albeit weakly—because the forces exerted by the phonons overcome the electrons' natural repulsion.

When the electrons pair, they don't bind to one another as though glued but, rather, travel along in unison, separated by some distance but moving together like surfers riding the same wave. Each has its own velocity, it is true, but each is able to change its velocity and, in so doing, changes its partner's as well. It is almost as if the moving electrons were attached to one another by a string or a spring, since they seem to snap each other along as they flow through the lattice of the conductor.

Since the lattice has been cooled, it does not vibrate as much, and the paired electrons, helping one another, glide effortlessly down the relatively still wire, their shared power more than enough to overcome any obstacles in the lattice. Their common momentum is totally unaffected by any random scattering of individual electrons that might occur, and electrical resistance is thus zero. Other electrons soon follow along in the wake, form pairs whose collective velocity is the same as all the other pairs, and the collective force of this pack is now the full superconducting current. It is the pairs, then, that are involved in the condensation of velocity Bardeen, Cooper, and Schrieffer had been seeking.

The electrons in a superconductor may be likened to a platoon of soldiers crossing a field strewn with ruts, holes, and rocks. If they travel singly, they might stumble and fall because of the obstacles. But by pairing off two by two, joining arms, and moving in close formation, they can proceed more smoothly: if one person falls, his partner can hold him up and keep him moving. The tightness of the formation helps, too. Since each pair always has another moving steadily in front and one in back, falling down is almost impossible.

Thus, the normally chaotic stream of electrons in a conventional conductor is tamed by the formation of these electron pairs. As Yale University physicist Daniel Prober explains it: "Independent electrons must occupy their own exclusive energy state, and they collide with electrons in other energy states and with defects in the conductor, giving off heat and dissipating energy. However, when they are linked together in pairs, they are all at the same energy level and move through the superconductors in an orderly fashion."

As long as the conductor is cooled to near absolute zero, the fragile electron pairs stay intact because cold reduces molecular motion. In that state, they may be likened to a trail of smoke in breezeless air. But even a bit of heat gives them enough energy to overcome the phonon attraction; atomic vibration in the lattice also becomes more violent, and the pairs break. As they break, superconductivity vanishes.

By 1959, the theory of superconductivity was firmly established, and even though there have been modifications, it has stood the test of time. But only, apparently, insofar as it relates to the conventional, metallic, low-temperature superconducting materials. The new high-temperature superconductors, the ceramics, pose another set of questions and may require an entirely new theory.

3

THE QUEST HEATS UP

NOT LONG AFTER Kamerlingh Onnes's monumental discovery of superconductivity in mercury, the list of metals
that could be turned into superconductors began to expand.
Tin, known to the ancient Egyptians, who could have no
inkling of the powers embedded in its thin skin, superconducted at the transition temperature of 3.73° K. (The
temperature at which resistance suddenly drops to zero in
a conductor is variously referred to as the superconducting
critical temperature T_c, the superconducting transition
temperature, the transition temperature, or simply T_c.) Zinc
superconducted at 0.91° K, and even lead, long known to be
a poor conductor of electricity, became a superconductor
when it hit its T_c of 7.22° K. Experiments with that lowly
metal have, in fact, produced consistently dramatic results.
All one has to do is cool a lead ring with liquid helium, and
the lead will carry a steady current of hundreds of amperes
around and around for years without any power boost
whatsoever.

Kamerlingh Onnes was keenly aware of the potential
value of his discovery of superconductivity, especially as it

might apply to the construction of yet smaller, enormously powerful electromagnets for use in industry, and to reducing losses in electric power systems. But there were a couple of dampers.

First, it was far from easy to reach the low temperatures required, and it was difficult to work at those temperatures. Precise measurements at such cold temperatures were also not easily obtained. One way to make such measurements was to measure the known pressure of, say, a known quantity of hydrogen or helium, a tricky procedure since the interrelationship of temperature and pressure at temperatures close to absolute zero is rather capricious. Intricate electrical devices must also be used to measure resistance; changes in magnetic properties must be carefully monitored, as well as the speed of sound waves traveling through helium.

The problems in measuring such phenomena were not easily overcome. The reporting of scientific results by the news media, however, jump all too quickly from concept to successful experiment. The long and tedious hours spent by researchers in arriving at their conclusions are rarely, if ever, reflected. Kamerlingh Onnes's first experiments, for example, required decades just to build the equipment that could reach down close to the boiling point of liquid helium. There also was little liquid helium available in 1900. Even today, when it is obtained as a by-product of natural gas production (the United States produces about 90 percent of the world's supply, extracting it from gas wells in Texas and Wyoming), liquid helium requires a good deal of energy to produce, has to be confined in tightly insulated containers, is dangerous in untrained hands, and expensive (up to eleven dollars a gallon).

The other problem Kamerlingh Onnes faced was with the superconductors themselves. When a metal becomes superconducting, it produces, as do ordinary conductors, a magnetic field. Initially, when a superconductor is cooled below its transition temperature and a magnetic field is increased around it, the magnetic field remains around the superconductor. This can be easily demonstrated in the classroom by supercooling a bit of the superconducting material, laying it in a dish, and then placing a tiny magnet over it. The magnet will float almost magically in the chilled air above the superconductor because unlike poles repel, forcing the superconductor and the magnet to move away from one another.

Discovered in 1933 and known as the Meissner effect, this phenomenon, along with the absence of resistance, is proof that a material is superconducting. But, Onnes discovered, when a current sufficient to stimulate a fairly healthy magnetic field was sent through the superconducting materials, the field ultimately penetrated into the materials, destroying the superconducting effect. In fact, even weak magnetic fields of a few hundred gauss did the trick, which meant that the metal reverted back to its normally resistive state, no matter what the temperature was. The point at which superconductivity lessens is called the critical magnetic field, or H_c, with the maximum destructiveness occurring at the lowest temperatures.

Because an electromagnet needs an electrical current to generate its magnetic field—typically, such a magnet consists of conducting wire wrapped around an iron core, which strengthens the magnetic field when electricity is pumped through the coil—building an electromagnet with superconducting materials seemed an insurmountable chore. On the one hand, the superconductors were carrying a frictionless flow of electrons, but on the other, there was a limit to the amount of current that could be sent through them, since that current would generate enough of a magnetic field to quench their conducting advantage. This wouldn't be too much of a problem if only applications that generated fairly small fields were considered. But most heavy-duty magnets run in the 10,000- to 100,000-gauss range, and that was a problem.

Two things, then, had to be done before there could be any major application of the newfound superconducting phenomenon. One was that new materials with higher transition temperatures had to be found. These materials would superconduct in ranges that were considered technically and economically practical—say, at the boiling point of cheaper, easier-to-handle liquid nitrogen (−320.5° F, or 77° K) or even, as impossible as it seemed, at room temperature. The other was that the materials had to be able to carry enough current to make their use competitive with copper cable for most applications, and be able to withstand the high magnetic fields without losing their superconducting properties. (When talking about current, scientists refer to current density, which measures how much electricity a superconductor can bear. A critical current density of 10,000 amperes per square centimeter, for example, means that a superconducting cable with a cross section of 1 square centimeter could carry a maximum of 10,000 amperes before being quenched, an amount that would put it in copper's league. Most applications need a critical current density between 10,000 and 1,000,000 amperes per square centimeter. Besides being able to carry a high current density, practical superconductors need to function in magnetic fields stronger than 2 tesla, which is forty thousand times stronger than the earth's magnetic field.)

With such problems resolved, said the dreamers, powerful, smaller electromagnets could be built, devices that might one day make the vision of a perpetual motion machine a reality. If a superconducting wire could be wound into a closed coil, electricity could be fed into it and channeled along a continuous path, as was done in the lead superconductor mentioned earlier. The current could flow swiftly and unimpeded, without a power boost—millions of watts of steady electrical power have to be sent through conventional electromagnets to keep them fired up—and without the heavy, bulky iron core that ordinary electromagnets require to jack up the magnetic field.

So the hunt for new materials began, but this time the focus was not on lower temperatures, as Kamerlingh Onnes's had been, but on higher ones. It was not easy. The process required meticulous surveying of scores of elements and simple alloys, and even some nonmetallic systems, like aromatic compounds, a chemical grouping which occurs in coal-tar products, benzene, essential oils, balsams, and other natural substances that have an aromatic odor. The behavior of such compounds became the special interest of Fritz London, who was the first to speculate, in 1937, that supercurrents might exist in nonmetallic systems. London suggested, in fact, that the phenomenon of superconductivity existed in certain large organic molecules, such as proteins, and that it was the superconducting state that explained some of the molecules' unusual properties.

Years later, Stanford physicist William Little continued along that rather lonely path of interest, generating a considerable amount of experimentation—along with scant success and a good deal of ridicule—in an effort to prove his notion that certain organic molecular arrangements would produce superconductivity at room temperature. The new generation of superconducting materials has since proved that Little was not so far off base as his critics believed.

As the scientists surveyed more and more materials, they became aware—even though they could not yet predict whether a metal would be a superconductor—that superconductivity was not the relatively rare phenomenon it appeared to be. In fact, its absence was quite often the exception. For example, when a metal wasn't superconducting on its own, or under the influence of the usual laboratory coolants, application of high pressure was enough to make them superconduct. Other elements, not superconductors themselves, formed compounds that were.

Niobium, discovered in 1801 and formerly called columbium, was a special favorite. A soft, ductile, gray-blue metal, it is present in several minerals, forms a number of compounds and complexes, and is used to strengthen welded joints and certain steels. In 1941, scientists reported superconductivity in niobium nitride with a T_c around 16° K, fairly high compared to earlier superconductors. Progress slowed after that until 1953, when John Hulm, of the University of Chicago, put together a material called vanadium-3 silicon. (Vanadium is a silvery white metal used in alloy steels and reacts with nonmetals at high temperatures; silicon is the metalloid element used in semiconductors.) This substance became a superconductor when cooled at a slightly higher temperature than niobium nitride, around 17.5° K.

Soon after, Bernd Matthias, then with Bell Laboratories, concocted something called niobium-3 tin, which handled superconductivity at 18° K, a milestone of sorts but still enormously cold when one considers the temperature is equal to –427° F. It's difficult not to compare these early efforts—indeed, the current-day ones as well—to those of medieval alchemists searching for the lapis philosophorum, the philosopher's stone, that ill-defined soluble substance endowed with the power to change base metals into gold and silver, a stone capable of purging a metal of its impurities so that it could be turned into some precious substance, or one with some useful purpose. It was all a matter of try it and hope.

Still, no matter how researchers varied the ingredients and their amounts, or how many they laboriously mixed, pulverized, calcified, crystallized, and precipitated, many of the materials they came up with were not, despite their superconductivity, all that serviceable. Niobium-3 tin, for example, could support large electric currents and remain superconducting, even in intense magnetic fields. But although promising, it proved to be too brittle and, as it turned out, was never used as extensively as scientists had hoped.

Also, of course, there was always the icy grip of cumbersome liquid or compressed gaseous helium that the materials had to be held in if they were to superconduct. Try as they might, researchers could not get the transition temperature of all their materials up to easily manageable levels. By 1973, although several hundred materials were known to superconduct, the best that scientists were able to achieve was a T_c of 23.2° K (–418° F) with a compound of three parts niobium and one of germanium, the latter a hard metalloid with, ironically, semiconductor properties.

This record high still required either liquid helium or hydrogen for cooling.

Not that there hadn't been any grand claims. "Like other men," René J. Dubos once said, "scientists become deaf and blind to any argument or evidence that does not fit into the thought pattern which circumstances have led them to follow." And so along with the welter of experiment and observation came a few flutters of higher T_c and some enormous rushes to judgment. Most of the results, however, were transient and nonreproducible—or just false.

Physicist Donald U. Gubser, acting superintendent of the Material Science and Technology Division of the Naval Research Laboratory, in Annapolis, observes:

> I like to call it the rocky road to high temperature superconductivity. There were some very wild, unsubstantiated claims, which made you wonder whether these people knew what they were talking about. They ignored the definition of the science of superconductivity, and every time they'd see a little glitch they'd say, "That's it."
>
> The first really wild one came from R. A. Ogg, Jr., back in 1946, when he claimed that dilute alkali metal-liquid-ammonia solutions became superconducting near 185° K, above liquid nitrogen temperatures, if the solution was rapidly cooled. The thing was, Ogg ignored the fact that the resistance in his sample was still 16 ohms—and in my mind, the formula for a superconductor is R equals zero, resistance equals zero. He had 16 ohms resistance. It dropped from, say 1,000 to 16, and he concluded that that large a drop had to mean superconductivity, that the 16 was probably just an end effect since he didn't have a very large sample. The experiment was neither reproducible nor widely accepted by the scientific community.
>
> Years later, the Russians also claimed it, even though the resistance in their samples wasn't

zero. The system they used was, in some situations, a very good conductor, and at times appeared to exhibit superconductivity. But it was unstable, and although several groups tried to reproduce the results, there was no confirmation. I can go back to all the early reports of high-temperature superconductivity and find resistance of some sort, and you'd be surprised how many times that "R equals zero" is ignored.

Not surprisingly, the failure to reach high superconducting transition temperatures led to gloomy predictions that there was a temperature limit—somewhere between 30° K and 40° K—beyond which no material, no matter how artfully it was coaxed along in the laboratory, would be a superconductor. Even Bernd Matthias confessed to being affected by a limit of 18° K, the same temperature at which his niobium-3 tin became superconducting; this limit had been predicted earlier by other scientists and, for Matthias, "always constituted at least a nagging reminder in my search for higher superconducting transition temperatures." In 1970, when the transition temperature had reached 21° K, Matthias wrote: "What else have we done? In many discussions, I have tried to point out that the present theoretical attempts to raise the superconducting transition temperature are the opium in the real world of superconductivity, where the highest T_c is, at present and at best, 21 K. Unless we accept this fact and submit to a dose of reality, honest and not so honest speculations will persist until all that is left in this field will be these scientific opium addicts, dreaming and reading one another's absurdities in a blue haze."

4

THE MATERIALS
REVOLUTION

TWENTY-FIVE YEARS ago, Stanford's William Little startled the general public with his predictions of plastic materials that had no electrical resistance at high temperatures, room-temperature superconductors, flying carpets, superconducting skis, trains that levitated over tracks and glided smoothly along at 300 miles per hour, and frictionless electrical transmission lines.

He even came up with a theory to explain how such materials would work and suggested that superconductivity might even be possible at temperatures upward to 3,632° F. His notions were published in *Physics Review Letters*, one of the most respected physics journals, and in *Scientific American*, the forum for the communication of a scientist's work to other scientists of different disciplines.

Little's ideas were intriguing to anyone who delighted in science fiction, but a source of dismay for his scientific colleagues. Lamented the *New Scientist* in London: "It is highly disappointing that the possibility of a room-temperature superconductor has been removed. The technological usefulness of such a plastic material would have been immense."

The prestigious journal had been bombarded with critical letters, and Stanford got plenty of telephone calls from scientists curious to know why Stanford University employed such a crackbrain. It is quite a different story today. Physicists, fired by a renewed—almost hysterical—interest in superconductivity, are jacking up transition temperatures in a growing new family of superconducting materials with a speed that is phenomenal compared to the snail's pace of researchers in the sixty-two years from 1911 to 1973, when the T_c was raised from $4°$ K to only $23°$ K. As anyone who reads the newspapers knows by now, materials that superconduct at around $95°$ K, using liquid nitrogen as the refrigerant, have been discovered. And, at this writing, there are reports of superconductivity at an astonishing $500°$ K, or $440°$ F, a temperature substantially higher than the boiling point of water. If these experimental findings are true and reproducible, the materials used would be the first superconductors that could work at or above room temperature, the holy grail of the field.

"I find it very, very satisfying," Little said recently, "and also a little bewildering, that most of the people who were such outright critics of this, when presented with an experimental discovery, have taken another look at it, and I see them saying there's absolutely no reason why you can't get superconductivity at room temperature. For twenty-three years they were not only scientifically opposed to it, but they were openly hostile to it."

The situation today is vastly different. Consider *Time* magazine's report of a recent superconductivity meeting of the American Physical Society in New York City:

> They began lining up outside the New York Hilton's Sutton Ballroom at 5:30 in the afternoon; by the time the doors opened at 6:45, recalls physicist Randy Simon, a member of TRW's Space and Technology Group, "it was a little bit frightening. There was a surge forward, and I was in front. I

walked into the room, but it wasn't under my own power." Recalls William Little: "I've never seen anything like it. Physicists are a fairly quiet lot, so to see them elbowing and fighting each other to get into the room was truly remarkable." [The meeting] was so turbulent, so emotional and so joyous that the prestigious journal *Science* felt compelled to describe it as a "happening." AT&T Bell Laboratories physicist Michael Schluter went even further, calling it the "Woodstock of physics." Indeed, at times it resembled a rock concert more than a scientific conference. Three thousand physicists tried to jam themselves into less than half that number of seats set up in the ballroom; the rest either watched from outside on television monitors or, to the dismay of the local fire marshal, crowded the aisles. For nearly eight hours, until after 3 A.M., the assembled scientists listened intently to one five-minute presentation after another, often cheering the speakers enthusiastically. Many lingered until dawn, eagerly discussing what they had heard and seen.

The Physical Society, however, does not by any means have a corner on the superconductivity market. Because the subject, in both its theory and potential for practical application, cuts across a number of scientific disciplines, one can learn about the latest breakthrough from a number of sources. (The word *breakthrough* itself, which once sent shudders through researchers when it was belabored by science writers, is now heard more and more frequently from the scientists themselves.) Important new information is likely to be gleaned at meetings of the Materials Research Society, the American Chemical Society, the American Association for the Advancement of Science, the U.S. Department of Energy, the New England Governors Conference, the International Society for Optical Engineering, the Metallurgical Society, and the American Medical Association, to name but a few. And at some, as the mob scene at the Physical Society meeting demonstrates, there are stampedes for information.

The annual meeting of the Materials Research Society, an organization of technical professionals from a wide variety of scientific and engineering disciplines, is a prime place to catch most of the current action. Considered a snore for years by science reporters—and who could blame them, given the dubious news value of such presentations as "Precipitate and Defect Formation in High-Dose Oxygen Implanted SOI Material," or "Metastable Defect Reactions of Carbon Interstitials in Silicon"—the MRS sessions have become must coverage for newspaper and broadcast editors. For it is here that the new superstars of superconductivity hold court. No matter now that the language is confounding, for the press conferences, the photo opportunities, and the hastier private interviews over coffee breaks cut through the haze. Just walking through an MRS meeting gives one a sense of superconductivity's status these days and wipes away any shred of doubt about its exalted place among all the other mysterious supers of science—supersymmetry and superfluidity, superstrings, superspace, and supergravity.

Free shopping bag in hand, the observer picks his or her way back and forth between rows of tables stacked high with research papers and brochures in the foyer of Boston's Marriott Hotel, jostling elbow to elbow with other shoppers who hardly glance at the titles but are hungry enough for knowledge to snatch anything in sight. Stuff it all into the bag: *The Cambridge Report on Superconductivity*, its lead story proclaiming cockily, "Here Come the Americans," reassuring the fearful among us that although the Japanese, with their patent applications and new-product announcements, have held center stage in the world for HTSC supremacy, "American firms in high temperature superconductivity are challenging the Japanese juggernaut." *Superconductor Week, the Newsletter of Record in the Field of Superconductivity*, again an oasis of encouragement with an "exclusive" interview with an IBM scientist who reveals that his team has unearthed a "critical clue" as to how the new superconducting materials work ("Yeah, yeah," mutters an MRS guy nearby, "now I can upgrade my PC, right?"). *NBS Update*, from the National Bureau of Standards, announcing that a "deeper sputter etch may be necessary for superconductors having long exposure to air." *Superconductors Update* ("The Pressure Is On. Get a Head Start on Your Competitors by Subscribing"). Notice of the forthcoming International Superconductor Applications Convention in Los Angeles ("special reduced hotel room rates and discounted air fare from American Airlines"), one of at least sixty HTSC meetings, workshops, short and long courses, scheduled in just the next two months. A reprint from *Business Week* ("Venture Capital's New Gold Rush: Superconductors"). A press release from the National Science Foundation announcing that the agency has launched a new initiative to speed research on superconducting materials "and help out the U.S. in the forefront of what seems certain to develop into a highly competitive industry." More reassurance, from the Department of Energy, quoting President Reagan: "Only a half year ago, superconductivity was considered a scientific backwater, a phenomenon with little practical purpose. Now scientists are saying it may change our lives. It shows all the dreams we have had can come true. The sky is the limit!" *High T_c Update*, which claims, "Our electronic mail distribution system has been unable to keep up with the demands being placed on it."

Advertisements that would have had Kamerlingh Onnes salivating offer further insight into the scope of current research:

YOUR CRYOGENIC CONNECTION
Cryosystems closed cycle turnkey refrigeration systems are ideal for characterizing the revolutionary new high temperature superconductors! Join the Race . . . Superconductivity at 28 K, 36 K, 39 K, 40 K, 70 K, 90 K??

OPPOSITES ARE ATTRACTED TO OXFORD
SUPERCONDUCTING MAGNETS &
SYSTEMS

Hey, hey. Split pairs to 10 Tesla . . . RT or cold bores, vertical or horizontal splits, top loading into vacuum, liquid or gas, standard optical systems for IR & UV. What more could you ask?

Well, you could glance at the bulletin board:

CALL FOR PAPERS: Original papers which have not been previously published are sought.

JOB POSITION, THE AEROSPACE CORPORA-TION: Qualifications: Ph.D. in solid-state phys-ics, experience with energy band calculation and/ or superlattice theory and/or electron transport. U.S. citizenship *required*.

MST, please join us in the MagLab tonight at MIT for a Scotch-filled evening of ion implanta-tion, the Raman Spectrum, and polarons in a non-parabolic band.

Or pick up snatches of conversation: "Listen, chemistry and biology make life better, but if you want to get rich, you have to go back to physics, and I heard that one, pal, from a Russian in Trieste." "Well, trying to figure out what they resemble, really, is difficult. They're sort of, well, like they look like they belong in a pottery shed, or maybe in a microwave oven, you know what I'm saying?"

Or watch the Japanese in uniform (blue serge suits, painfully starched white shirts, blue polyester ties, brown slip-on shoes with fake laces), in pairs, taking notes, pic-tures, and chances (they're standing outside a meeting room near which someone has stuck up a hand-lettered sign, "English only spoken here").

Or listen to a paper or two (on the theory that a person can learn only a half dozen or so new words a day when studying a foreign language, one presentation is probably enough for the uninitiated), like Paper 15.2/L4.2, on "Characteristics of Polyethylene Thin Films Deposited by Ionized Cluster Beam." (For the faint of heart, Dr. William DeVries from Humana Hospital was giving a nonsuperconductivity plenary address on the artificial heart at this particular meeting.)

Such enormous interest in superconductivity ("The theorists are having a ball, the experimentalists are having a ball, and next week alone, I'm giving four talks on this thing, three of them on different aspects," exulted the Naval Research Lab's Don Gubser) is not confined to the scientists, or to such esoteric journals as *Low Temperature Physics*. Financial media like *Business Week*, *The Wall Street Journal*, *Fortune*, *Money*, and *Forbes* have become frequent forums as the venture capitalists—lured by the promise of ultrafast computers, supermagnets, high-speed train travel, fusion reactors, power transmission without loss, efficient energy-storage systems, medical diagnostic equipment—go for the jackpot: the new multibillion-dollar industries of the 1990s.

Companies have sprung up like weeds as the search for the right stuff heats up. Their names are as bold as their mission: Superconductivity, Inc.; Monolithic Superconductors, Inc.; American Superconductor Corp. Many are wooing the Defense Department in a bid for research funds to develop superconducting wire and thin films of material for integrated circuits in computers. The department plans to spend $150 million on superconductors over the next couple of years, backing everything from the Star Wars defense system to communications, even though the technology required to make something of the new materials—and the basic science behind superconductivity itself—still have a way to go.

Even an amateur without big bucks can break into the field—with a kit demonstrating the Meissner effect of magnetic levitation. For twenty-five dollars, the Institute for Chemical Education at the University of Wisconsin will send superconductor buffs a kit that includes a 1-inch pellet of a new superconducting compound, special magnets, and detailed instructions for performing levitation experiments (liquid nitrogen, container, insulated rubber-coated gloves, and safety goggles not included).

What is perhaps most remarkable about the recent superconductivity craze, however, is not so much that superconductivity could be achieved at temperatures higher than the 23° K that stopped researchers short for so many years, but that it was so unexpected in the sort of materials now under scrutiny. Though the metallic and intermetallic compounds that had been tested had proven themselves, the temperatures required to make them superconduct were appallingly low. It was beginning to look as though the pessimists were right, that there was a limit to what could be accomplished. Fortunately, this was not to be the case. While most scientists abandoned the quest, a handful hung on, driven by the belief that maybe, just maybe, there were unexplored materials that might be the answer, even a new science that would spring from the ashes to justify them.

There already was precedent for such thinking. As early as 1965, William Little was speculating in *Scientific American* about the "possibility of discovering some other substance—perhaps a nonmetallic one—that would be superconducting at higher temperatures":

> As a matter of fact, it is an especially opportune time to investigate such a possibility in view of the great theoretical advances that have been made in recent years toward understanding the superconducting state. I have been particularly interested in the possibility of synthesizing an organic substance that would mimic the essential properties of a superconducting metal. My calculations have shown that certain organic molecules should be able to exist in the superconducting state at temperatures as high as room temperature [about 300° K] and perhaps even higher!

Little's interest in the possibility of biological superconductivity had been stimulated five years earlier while he
was working at Stanford on the problem of heat transfer to
a metallic superconductor.

> I was struck by the great stability of the super-
> conducting state, and it occurred to me that if
> nature wanted to protect the information con-
> tained, say, in the genetic code of a species
> against the ravages of heat and other external
> influences, the principle of superconductivity
> would be well suited for the purpose.
>
> In view of nature's remarkable record of inge-
> nuity in such matters, I thought it might be use-
> ful to determine if the superconducting state
> could occur in a large organic molecule built
> along the general lines of the genetic molecule,
> DNA [deoxyribonucleic acid], the carrier of ge-
> netic information in all living cells.

Little's notion that it might be possible to synthesize an
organic superconductor envisioned an organic polymer, a
substance with large molecules, doing the job. Synthetic
polymers, widely used in plastics, are made by linking a
large number of identical chemical building blocks, called
monomers, into long molecular chains. The first material of
this type, celluloid, was created in 1868.

There are also many natural polymers, and among these
are the polysaccharides, complex carbohydrates made up
of simple-sugar building blocks tied together in long
strings. Starches are polysaccharides, as are glycogen, the
main source of carbohydrate energy in animal cells, and
cellulose, which supplies the support structure in plants.

Little's superconducting polymer would have long molecules and an atomic structure similar to a skeleton. Two
electrons would move down the backbone of the structure;
at regular intervals they would pass ribs, or chains of
complex atomic structures, shoving other electrons out of
the way along the ribs. The positive field created by the

departing electrons would attract another electron in the spine to form the pair needed for the BCS theory, the beginning of the action required to produce superconductivity.

But while the BCS theory applied to low-temperature superconductivity, Little's model, after detailed calculations were carried out, had an enormously high transition temperature—typically around 2,000° K.

Years later, a few scientists intrigued with Little's theory were predicting room-temperature superconductivity in DNA molecules themselves, and at 140° K in such oddments as bile salts. Again, there was not much in the way of confirmation, and reports of very high temperature superconductivity in organic materials began to fade. But in 1975 superconductivity showed up in a polymeric material, and although the temperature was low, below 1° K, the discovery demonstrated that the phenomenon of superconductivity was not restricted to the conventional alloy systems. Eventually, aided by advances in organic synthetic chemistry, new organic superconductors were discovered, and T_c began to climb, although, again, not to very high readings.

Then, in the early 1970s, came a discovery that, for people accustomed to viewing plastics as insulators, seemed bizarre: polymers that apparently handled normal electrical conductivity as efficiently as copper. Like Luigi Galvani's experiments, the discovery was accidental. An account in *Scientific American*, tells how it came about:

> A graduate student in Hideki Shirakawa's laboratory at the Tokyo Institute of Technology was trying to make a polymer called polyacetylene from ordinary acetylene welding gas. The polymer, a dark powder, had first been synthesized in 1955, but no one knew much about it. Instead of a dark powder, Shirakawa's student produced a lustrous, silvery film that looked like aluminum foil but stretched like Saran wrap. Looking back over his chemical recipe, the student saw his

mistake: he had added 1,000 times more catalyst than the instructions called for. What he had made was polyacetylene, all right, but in a form different from any polyacetylene before it. . . . The search for so-called synthetic metals had just begun, and in its new guise polyacetylene was an obvious target for investigation. Shirakawa subsequently spent a year at the University of Pennsylvania . . . exploring the potential of the revamped polymer. The collaboration bore fruit when investigators tried doping the material with iodine. [Doping refers to the process of incorporating small quantities of certain chemicals into another substance. To dope a polymer, it must be saturated with chemicals that cause an excess of positive or negative electrical charge to build up along the backbone of the polymer chain. Doping, as we will soon see, played a key role in the discovery of the new superconducting materials.] The flexible silver films became metallic golden sheets and the conductivity of polyacetylene increased by more than a billion times!

Since then, about a dozen polymers and polymer derivatives have undergone the transition when doped. Today, in fact, conducting polymers match the conductivity of copper, and only recently the first rechargeable polymer battery was marketed.

But conducting plastics were not the only oddball substances that had started to pique the curiosity of researchers hunting for better ways to transport electrical current, not to mention trying to surpass the 23° K T_c that seemed as impassable as the sound barrier had been to fast flight. There was another group of materials still to be looked at, the mixed metallic oxides: substances with the mechanical and physical properties of ceramics. They weren't exactly the sort of thing one might expect to find at the center of attention in a modern physics laboratory.

Ceramic materials, after all, seemed more at home in a pottery class or factory. The stuff of bricks and drainpipes, of bathtubs, sinks, and toilet bowls, of cups and plates, these earthy compounds, especially clay, were useful but too ordinary and familiar to interest Ph.D.s tinkering about with differential scanning calorimetry instruments or scrutinizing the partial penetration of flux. Furthermore, the majority of ceramics are insulators, not conductors, and are so used on high-voltage electrical transmission lines.

But not all ceramic materials are clay (glass is made of sand), and they don't all behave the same way. They are, it turns out, far more interesting and versatile than one might imagine. For years, ceramics were generally defined as inorganic, nonmetallic materials, are known primarily as silicate bearing, that is, composed of the natural minerals that make up most of the rocks in the earth's crust.

Among the silicates are clay, bauxite, feldspar, silica, and talc—the stuff from which a horde of valuable products and materials are made. But it's important to know that most ceramic materials now include many compounds of carbon, nitrogen, and oxygen combined with other elements: carbides, silicides, nitrides, and metallic oxides, all of which have various electrical properties. Lastly, there are cermets, various combinations of ceramics and metals, with rather complex molecular structures and the ability to withstand extreme temperature and resist corrosion or abrasive wear. And so we hear today of high-technology ceramics, advanced ceramics that find their way into turbocharger rotors, the garnets in lasers, the guts of computer memories, space shuttle tiles, and engines. Ceramic engineers today carefully control the proportion and type of materials and, unlike the way traditional ceramics are prepared from abundant raw materials with simple processing techniques, the manufacture of, say, electronic ceramics requires precise control of the microstructure and composition of the very grains that comprise a given substance.

Thus, the scientists who were starting to look more carefully at ceramic materials for new superconductors were making a sharp distinction between what lay people know as ceramics—the word comes from the Greek *keramos*, which means burnt stuff—and a more sophisticated group known as the rare and alkaline earths. (Strictly speaking, the "earths" are the oxides of elements, and the so-called rare ones are really ubiquitous, having acquired their name because they were first isolated in oxide form from rare minerals.) Although what the researchers had been doing all along smacked of alchemy, no one was about to simply draw out a soggy lump of clay into a wire, send an electrical current through it, and expect it to superconduct.

Among the many earths, there are three that figure prominently in the development of high-temperature superconductivity. Barium, a soft, silver-white metallic element of the alkaline-earth group, occurs only in combination and, because it is opaque to x-ray, is known to all those sufferers of stomach and intestinal disorders who have had to undergo diagnostic barium enemas or swallows. Another silvery metallic element is lanthanum, a rare earth used as a catalyst in cracking crude oil, and as a component in cigarette-lighter flints and, in oxide forms, in optical glasses. The third is yttrium, gray and metallic, used in alloys to make strong, permanent magnets and in color television phosphors to provide luminescence. It was these three—barium, lanthanum, and yttrium—along with the combination of copper and oxygen known as copper oxide, that would bring superconductors in from the cold.

Although a number of scientists felt strongly that it was a waste of time to test materials that were not related molecularly to known superconductors, a few held to the belief that these nonconductors might be coaxed into behaving differently by systematically altering the proportions of ingredients in ceramic materials and by modifying the all-important crystal structure under the right temperature and pressure. They believed that by harnessing the mighty atomic forces that held the ceramics together like concrete, they could get electrons in these "dead" substances to loosen their bond to the atomic nuclei and jump from atom to atom, as they do in ordinary good conductors. Thus they might achieve the unthinkable—the transformation of an insulator into not only a conductor, but something far more awesome. Like the bumblebee—which, aerodynamics tells us, should not fly—their "nonconductors" would fly in the face of commonly accepted beliefs.

There had already been some modest successes with superconducting oxides. Among these was strontium titanate, the first oxide superconductor and the first so-called perovskite superconducting material. (Perovskites, a large group of metal oxides, are natural minerals with a three-dimensional crystal structure.) Although the T_c of the material was extremely low, around $1°$ K, the achievement was still a major milestone and spurred continuing, if sporadic, research into the oxides. This resulted in the discovery, in 1975, of superconductivity in a mixture of lead, bismuth, and barium oxide at a T_c of $14°$ K. Though the T_c was still frigid, it nonetheless dispelled the widely held belief that superconductivity in the oxides was limited to very low temperatures.

In the early 1980s, a pair of French crystallographers at the University of Caen planted a seed that, in 1986, would start the superconductivity revolution in earnest. Claude Michel and Bernard Raveau had been testing a ceramic material made up of copper oxide laced with small amounts of barium and lanthanum. Interested in the mixture as a possible catalyst to speed up chemical reactions, they didn't test it for superconductivity but did note that it seemed to be a pretty good conductor at room temperature. No one paid too much attention to the reports at the time, but in 1983 J. Georg Bednorz and K. Alex Müller of the IBM Research Laboratory near Zurich got interested. Müller, Swiss-born, then fifty-five, was a solid-state physicist with a dogged personality and, as an IBM fellow, the time to pursue his own research interests. According to some reports, he did most of his heavy thinking, not in the laboratory, but on skiing and hiking trips in the Swiss Alps. Totally fascinated by the prospect of superconductivity, even though he hadn't been working in the field, Müller had a hunch that certain oxides might somehow march to a different drummer and, instead of insulating, superconduct. But before he could look for the right materials, he had to have help from someone more tuned to the laboratory than he was. He selected Bednorz, then thirty-two, a meticulous experimenter from Germany who joined IBM only the year before, and an expert grower of crystals that carry electrical charges.

Between them, Bednorz and Müller surveyed hundreds of different oxide compounds, searching for those with a high concentration of electrons with very strong attractive or coupling forces between pairs—the requisite for superconductivity. They suspected that the right oxides had to be out there somewhere, oxides that might have properties not entirely covered by the prevailing theories of superconductivity. They even guessed that oxides containing copper might demonstrate critical temperatures higher than the theories predicted. It was a shake-and-bake operation. Techniques more often used in a ceramist's workshop than in a high-tech IBM lab began to take center stage, and terms such as *annealing* (heat treatment to soften a substance) and *sintering* (heating and compacting a powdered material below its melting point to weld the fine particles together into coarse lumps) became counterpoint to *thermogravimetric analysis* and *x-ray diffraction*.

Instead of trying to fashion the preparations as crystals, which involves a far more demanding process and results in materials that would have been most difficult to handle, Bednorz ground, mixed, and fired the preparations as a potter would his or her clays, with simple mortar, pestle, and oven. The two researchers worked quietly, relying purely on their intuition, with the conviction that they had little to lose by their trial and error method, given that other researchers had not accomplished all that much either. "We didn't tell anybody what we were doing because we would have had difficulty convincing our colleagues that what we were doing wasn't crazy," said Bednorz later.

Toward the end of 1985, they had little to show for all of

their powdering, pressing, molding, and baking—until Müller, going through the literature in his library, happened on the report by Michel and Raveau. The copper in the mix seemed to be exactly what he and Bednorz had been searching for. Like jazz pianists varying a familiar theme without benefit of a score, they put chemical notes in places where they had not been before. Change the ratios of copper valence states here, alter the oxygen content of the sample there, substitute some barium for some lanthanum, substitute some more, heat it and cool it down, and check, check, and check again for evidence of superconductivity in the tiny crystals, the minuscule grains, that made up the ceramic material.

In January 1986, Bednorz was working alone in his lab with his latest mix. Instead of just combining the powdery oxides, as the French team had done, he dissolved them in water, allowed the particles to settle, then fired them at around 1,800° F. What he came up with was a sandwich consisting of layers of lanthanum and barium atoms alternating with layers of copper and oxygen atoms.

The odds against making an entirely new superconductor from scratch without even an applicable theory to follow are not encouraging. The number of possible combinations alone is staggering. Processing conditions vary markedly from lab to lab and can have an enormous effect on the structural properties of a sample. The level of barium-doping must be exact. The right microscopic configuration of copper and oxygen atoms is crucial and depends on the right annealing and sintering temperatures, as well as careful cooling. Given all these variables, what happened was—to use an otherwise extravagant term—truly monumental. When Bednorz shot a current through his sample—it had now been chilled with liquid helium—it superconducted at −396.4° F, or 35° K, a startling 12° K above the old record set back in 1973, and 21° K over the previous record for a metal-oxide superconductor.

For years, researchers had been putting in just as many long hours, working with all manner of materials, but had managed temperature leaps of only a few degrees before hitting a barrier at 23° K. According to some theoreticians, there was no way to go beyond. If the past was any guide to the future, what Bednorz and Müller had done should not have been accomplished until the next century. In October 1987, just twenty-one months after their breakthrough, the two scientists were chosen to receive the Nobel Prize in physics, unprecedented speed for conferring the award, but a forecast of the pace of the discoveries that would follow over only the next few months.

In the weeks after the Nobel was awarded, newspaper and magazine readers would be looking at full-page advertisements from IBM, trumpeting the discovery of a whole new class of superconducting materials under a most unusual headline that, in a few strokes of scientific jargon, said it all:

The now-famous formula for the first superconducting ceramic: lanthanum-barium-copper-oxide.

It was the formula for the ceramic superconductor, and under it appeared the words, "Who knows where it will stop?"

Despite their elation, though, Bednorz and Müller were not especially eager to broadcast their findings. For one thing, they had not yet done the crucial Meissner repulsion test, the best proof of superconductivity, because the magnetometer they needed to do it had yet to arrive from the manufacturer. For another, they were quite aware that, given the slow progress of earlier work, the temperature jump over the old record would be received with skepticism by some of their colleagues. So they wrote a cautious paper and, to cover themselves and at least get on the record, submitted it in April 1986 to a German science journal, *Zeitschrift fur Physik*, once widely read but at the time not exactly the forum for pioneering research. It took six months for the article to appear, and when it did, the doubters surfaced on schedule. "Our group read the paper," said physicist Douglas Finnemore at Iowa State University. "We held a meeting and decided there was nothing to it."

Not everyone shared that pessimism, however, least of all Japanese scientists. Not generally known for doing science merely for its own sake, they invariably have their antennae out for any new bit of research that would have—to use one of their favorite phrases—"an epochal influence on the human race in the twenty-first century," which is another way of saying "any new research that might have strong commercial possibilities." A recent Japanese government proposal for an international effort to "explore the frontiers of science" drew complaints from researchers in a number of countries. Once again, they said, the Japanese, who spend very little on basic research, were striving to gain better access to advanced technologies by getting others to do the dirty work, and were merely picking the world's brains. (Interestingly, Japan has a research facility with the embarrassing sobriquet, City of Brains, at Tsukuba, north of Tokyo. Begun in 1966 as a sort of astronomy-to-zoology Silicon Valley, its atmosphere was designed to stimulate independent thinking and was Japan's answer to charges that its scientists neglect basic science. Recently, as we'll see later, it was the site of some important research in ceramic superconductivity.)

But while Japanese scientists have given relatively short shrift to the more brainy aspects of their trade (Japan won its first Nobel Prize in chemistry in 1981 and has three in physics), aggressive and highly coordinated programs using ceramics have been underway for several years, notably to develop ceramic engines that would give cars of the future durability, high acceleration, corrosion resistance, and fuel economy. And although superconductivity research had slowed over the years, as it had throughout the world, it had never stopped completely. Japanese scientists had already turned out high-performance superconducting wire using conventional materials (some of it used in a reactor at the Lawrence Livermore Laboratory in California), and they had chaired several international conferences dealing with the possibility of finding new superconducting materials among organics, alloys, and ceramics.

So, when Müller and Bednorz's report came out in *Zeitschrift fur Physik,* Japanese physicists not only saw the obscure German journal, but read it very carefully. At first, there was a touch of skepticism, but just a touch. One specialist in superconductors, Koichi Kitazawa, a member of Shoji Tanaka's team at Tokyo University, was not immediately impressed; he tossed it off to a graduate student in physics who had asked for permission to repeat the IBM scientists' experiment, as one account of the incident had it, and "make a joke of it." By November 1986, the Japanese weren't laughing: they had done the Meissner test, and the ceramic responded magnetically, confirmation that the Müller-Bednorz discovery was no fluke.

The next month, at a meeting of the Materials Research Society in Boston, a lot more scientists were going to know about it. The symposium on superconductivity, which had been routinely scheduled, was dragging on—until the fourth day, when the scientists were just about ready to pack it in. That afternoon, facing an audience of dozing and only half-interested researchers, Chinese-born physicist Paul Ching-Wu Chu, from an obscure lab at the University of Houston, unprepossessing at five foot six and hardly a household name in the research world, got up to deliver his presentation. Speaking in accented tones that seemed certain to induce full-scale narcolepsy, Chu was midway through his paper when he noted, almost in passing, that his laboratory had repeated Bednorz and Müller's experiment—and had verified the resistance drop. Kitazawa was in the audience, and immediately after Chu finished, he was on his feet, excited and clutching a pair of Vu-Graphs in his hand. He came to the podium and screened them, not only confirming the drop in resistance but adding in the vital part about the Meissner effect. Later, Gordon Pike of the Sandia National Laboratories in Albuquerque commented, "I think he'd been planning to keep the findings under the table for a while, but Chu forced his hand."

The report had a catalyzing effect. What had been a ho-hum field of research was now the scene of frenzied world-wide competition as the hunt for higher-temperature superconductors was on once again. Only this time the search had a narrower focus, the oxides. Among the first to jump into the race was Bell Communications Research (Bellcore), which provides research and technical support for the telephone companies formed in the breakup of the Bell System.

Bellcore's keen interest was understandable. The central office of a telephone company is, after all, one gigantic computer, and the new materials offered endless opportunities to improve communications technology. For example, very large-scale integrated circuitry (known as VLSI in the trade), the tiny complex of electronic components and their connections that are embedded in or on thin slices of silicon—the familiar microchips—could benefit enormously from the development of practical, high-temperature superconducting materials.

Progress in VLSI had been rapidly approaching an impasse due largely to the heat created by electrical resistance in the fine wires that interconnect both the microchips and the tens of thousands of tiny electronic devices jam-packed onto each individual chip. But superconducting interconnections would have no resistance to electricity and would generate no heat. Electronic devices could, therefore, be brought into much closer proximity, permitting even larger scale integration, thus opening the door to a new generation of ultrafast computers.

Bellcore, like all the other groups that had just burst out of the starting gate, concentrated at first on replicating the results of IBM's Zurich team. Its researchers fabricated a similar compound and managed a modest gain, to 38° K. Ironically, however, they had put together something else—an oxide of yttrium, barium, and copper—that was to cause them the same regret as that often expressed by entrepreneurs who decided not to buy Xerox stock when it was offered at a song. Tests on samples of this oxide had shown little promise, so the entire batch was put on the shelf in favor of more likely compounds.

Not so with Paul Chu. No novice in the field, he had been working with superconductivity since 1965 (originally at Bell Labs), toiling seven days a week in his grubby, crowded "Third World Lab," as he called it, and in a closet-sized office carpeted with scientific papers. An aggressive researcher willing and eager to press his luck to the limit, yet a perfectionist who knew that luck alone was not enough, Chu drove his tight little staff hard, with an enthusiasm that was infectious. "We are all friends here," said Ru Ling Meng, an associate whom Chu had recruited in the late seventies from mainland China. "When I am away for more than a day, I miss the place. All of us are attracted by Chu's character. The time spent here is very efficient. You feel so satisfied."

Chu had tested a vast range of potential superconductors over the years, and though disappointed when they all failed to meet his expectations, he was never fearful of failure or irritated at the time spent. "We were willing to try all kinds of wild experiments," he said, "to get the temperature higher. We never got ourselves boxed into conventional thinking." It was Chu's propensity for thinking wildly that finally paid off.

It had long been known that ultrahigh pressures could pack the atoms and molecules of some substances into more substantial configurations. Gases, as we have already noted, could be liquefied by putting them under pressure, and some metals that did not superconduct on their own eventually did so when pressure was applied. Among the pioneers in the field was Percy Williams Bridgman, an American physicist at Harvard University whose contributions resulted in new techniques that increased laboratory pressures nearly a hundredfold, and won him a Nobel Prize in 1946. Once, when he reached a pressure twenty thousand times that of normal atmospheric pressure—physicists use the term *20,000 atmospheres*—he burst the metal containers used in the experiment. Eventually, he developed more resilient materials, and succeeded in creating pressures of over 1,000,000 atmospheres. Bridgman also got ordinary yellow phosphorus, a nonconductor, to conduct electricity after transforming it under pressure into a black chunk of material.

Curious about what would happen to the IBM ceramic if pressure were applied, Chu put some of it under a crushing 12,000 atmospheres—and got it to superconduct at around 52° K (−366° F), a sizable boost over what others had managed, but still far too cold to raise much of a commotion among his fellow scientists. Chu tried more pressure, but nothing happened. In fact, a Bellcore team had gotten somewhat similar results without pressure. So Chu decided to pack his ceramic's molecules in a different way: he doped the lanthanum-barium–copper oxide with strontium, an alkaline-earth element similar chemically to barium but with a smaller atomic structure, and made a first-generation superconductor that did the job at 54° K (−362.2° F), another record high. Since the strontium's smaller atoms seemed to be doing something, Chu tried calcium, a very common element with even tinier atoms. He got an opposite reaction: the temperature dropped.

Chu kept it up, trying one recipe after another like a chef in pursuit of the perfect sauce. Finally, he and his team, along with Maw-Kuen Wu at the team's University of Alabama unit (Wu was a former graduate student of Chu's), replaced the lanthanum with the rare earth yttrium. They heated the mixture for hours at 1,652° F, ground the solid mass produced, and sintered it at 2,192° F. Wu drenched it with coolant, but this time he used liquid nitrogen. When Wu passed an electric current through the new ceramic, its resistance dropped sharply—at what physicists would later call "a balmy" 93° K (−292° F). "We were so excited and so nervous," recalled Wu, "that our hands were shaking. At first we were suspicious that it was an error."

It was not. A few days later, he and Chu did it again, bettering the mark by five degrees. The standard was now an impressive 98° K (−283° F), and liquid nitrogen had another role to play, perhaps the most important in its extensive repertoire. The discovery meant that expensive and volatile liquid helium cooling could now be shelved, or at least deprived of its exclusivity when it came to putting superconductivity to work. There was also a new formula on the block:

$$YBa_2Cu_3O_{7-x}$$

or as it was nicknamed, the 1-2-3 compound, for the ratio of its metal ions. (The subscripts in these formulae represent the number of atoms of each element in the unit crystal, the smallest atomic arrangement that describes the basic structure. The "7–x" subscript for oxygen indicates that oxygen content varies with preparation conditions but averages slightly less than seven atoms. Typical values range from 6.8 to 6.9, and we'll see later just how important oxygen is to superconductivity.)

The discoveries of Chu and Wu might seem trivial by a layperson's standards. After all, −283° F was not exactly beach weather. But it has to be kept in mind that superconductivity was originally observed only at temperatures around −460° F. The difference between the two, from a scientist's standpoint, was towering. Donna Fitzpatrick, assistant secretary for conservation and renewable energy at the U.S. Department of Energy, said shortly after Chu's discovery:

> Nobody ever thought ceramics would do this. Why should a ceramic conduct? That was the most interesting part of the whole thing. We had no theory that predicted this, and no theory to explain it. I think everybody agrees that this is a once-in-a-lifetime event, and we're all very excited about it. And even though most people will tell you that we won't have something we can use tomorrow—and that maybe we'll get to room temperature and maybe we won't—that jump into the nineties was astonishing. What really excites people is the theoretical question mark. We have something that we can't explain, and when we do, where is it all going to take us?

As astonishing as it was, however, the 1-2-3 formulation was not to be the end of the line. The scientists were not finished playing their game of chemical Scrabble simply because they had spelled out nitrogen and messed up helium's place on the board. Ready to pull a lot of late-nighters, they moved into their labs full-time, catching naps on mattresses laid on the floor, snatching meals of junk food at their workbenches.

Scientists in Japan and China, working independently, had already raised the transition temperature of the 1-2-3 compound before any of Chu's publications were released. And a horde of other researchers quickly duplicated those results, along with Chu's. At Bellcore, after the Chinese announced their results, researchers quickly confirmed it; a day or two later, they took down their shelved oxide and found, to their chagrin, that it contained the same ingredients that Chu and Wu had used. At the Brookhaven National Laboratory on Long Island, New York, metallurgists hit 90° K by substituting the rare earth lutetium—which up till then had no known uses—for some of the lanthanum; at IBM's Almaden Research Laboratory in San Jose, California, scientists also got to 90° K with another knockoff—replacing the yttrium with a mixed bag of chemical jawbreakers including dysprosium, ytterbium, samarium, and gadolinium.

From then on, scientists, Chu among them, deluged their favorite journals and program chairmen with reports of higher and higher transition temperatures. Universities and manufacturers of scientific equipment spat out news releases. Everyone, it seemed, either was now on the verge of a room-temperature superconductor or had seen it happen, if only momentarily; or at the very least, they had gotten into the dry-ice range, or that of a home refrigerator, or as one physicist put it, that of a cold day in Alaska.

The Chinese People's Daily trumpeted 100° K, then 215° K (admitting that the transition was very broad, with a midpoint at 93° K. An IBM team, working with samples of 1-2-3 at higher than normal temperatures (up to 2,192° F, versus 1,652° F) measured a resistance drop around

200° K; later they determined that the drop was a fluke, that subtle shifts in resistance in the contacts between the electrical leads and the sample, and not in the sample itself, were responsible. Sumitomo Electric Industries of Japan came in with 300° K (no confirmation). In Michigan, researchers at Energy Conversion Devices announced that part of a synthetic material made of fluorine (a highly dangerous yellow gas), yttrium, barium, and copper oxide had superconducted at 45° to 90° F. (The "part that superconducted, it turned out, represented less than 1 percent of the material tested, and the samples were far too small to lose all resistance. It is incredibly difficult to identify the exact portion of any material that shows superconductivity and then produce a pure sample of it.) In New Delhi, at the National Physical Laboratory, scientists saw evidence of superconductivity in material heated to 80° F, but the electrical signals were misleading, an artifact of the measurement process.

The flurry of reports prompted Jerry B. Torrance of the IBM research team to liken the situation to a TV game show whose contestants, in their haste to be the first to answer the prize question, were sounding the buzzer before they were sure of the answer.

The highest T_c of all (at this writing) was the 500° K (440° F) observed by Ahmet Erbil, a Georgia Tech physicist. Working with systematically altered samples of the 1-2-3 compound (which, incidentally, is both stable and easily reproducible), Erbil found one that seemed to superconduct at the astonishingly high temperature. The compound appeared to be stable—that is, when high heat was applied, it did not lose its ability to superconduct—and Erbil was able to make several different samples, all of which worked.

"At this point, I don't know whether it's real or not," Robert C. Dynes, of AT&T Bell Laboratories in Murray Hill, New Jersey, said at the time. "I've seen the data. Obviously, there are some questions that [the researcher] doesn't know the answers to, and so it's got to be aired out. But it's not nonsense."

Erbil himself was restrained. "This is a scientific break-through," he said, "but it is not a technological one yet in the sense that we need to improve the materials. It is revolutionary because of its potential."

Chu, as well, had seen flickers of superconductivity in temperature ranges that had come in from the cold. In June of 1987, he reported that one of his mixes reached a transition temperature of 300° K, or 78° F, but that it was unstable—meaning that when higher heat was applied to the material, it lost its ability to superconduct, perhaps because the added heat damages the chemical bonds that link the atoms.

In February of 1988, however, Chu—and researchers working on their own at the National Institute for Metals at Japan's City of Brains, at Tsukuba—came up with something that finally seemed legitimate. The compounds used by the two groups were very similar, mixtures of bismuth—a crystalline metal used to make alloys, heat-activated safety devices for fire detection and sprinkler systems, and medical and cosmetic preparations—strontium, calcium, copper, and oxygen (Chu's also contained aluminum). The Japanese ceramic had zero resistance at 105° K, Chu's at 114° K.

Almost as soon as the word of the new material was out, laboratory groups everywhere began mixing up batches, and they reproduced the results with lightning speed. It was new stuff, it showed promise of being more easily worked into useful materials than the other mixtures—it was less brittle than the others, which meant it might be more readily fashioned into wire—and it seemed to be able to carry larger currents at liquid nitrogen temperatures. It went nowhere near room temperature, but no one cared about that; it was real.

There was still more to come, and not only from the big, well-heeled companies and universities. At the University of Arkansas, Allen Hermann, chairman of the physics department, decided to go for a high T_c mark. Co-inventor of a heart pacemaker battery and a jazz trombone player of some note (stints with Ella Fitzgerald and Lionel Hampton), Hermann may not have seemed destined for great things in superconductivity. He didn't even have a proper lab. But Hermann, with the assistance of a visiting Chinese chemist, scrounged around for leftover chemicals and equipment and went to work. He seized on calcium and the metallic element thallium as additives and started mixing up a batch. Unbeknownst to Hermann, thallium was also deadly—it is used in rat poison—and the only ventilation in his lab was a few open windows. Warned of the danger by chemists, Hermann and his crew donned masks and gloves and moved into ventilated space. His efforts were successful: Hermann pushed the transition temperature higher still, to around 120° K.

Eighteen hours later, researchers at Du Pont Company had duplicated Hermann's work. Soon afterward, an IBM team came up with a version of the thallium superconductor that did the job at 125° K. Over at AT&T Bell Labs, a team hit the same temperature with a combination of barium, potassium, bismuth, and oxygen, the first of the new superconducting oxides to work without copper. Hermann was undaunted by all the upstaging. Not long ago, he reported on his thallium superconductor to a physics convention in New Orleans and left early to play at a local club. Said he: "I've got to keep my priorities in order." For the superconductor scientists, it was, indeed, beginning to look as though Andy Warhol was right: people are famous for only fifteen minutes.

5

GETTING DOWN
TO THE WIRE

NOT TOO LONG ago, a Japanese report on superconductivity research and development suggested that any technology that resulted from the new high-temperature ceramic materials should be regarded as international property. "Thus," the report went on, "it is important for Japan to bring out fully its accomplishments, and to contribute to the world through international cooperation. International cooperation will be stepped up through international symposiums and research collaboration."

A few months later, at a Boston meeting on superconductivity, a somewhat different response was indicated by Shoji Tanaka, who had led the University of Tokyo team that confirmed IBM's discovery of superconductivity in the lanthanum compound. Asked to comment on recent reports that the Japanese had made great progress in firing a ceramic into a serviceable wire, Tanaka said, "My work is not in that direction," adding, after a fairly long pause, "but in any event that might be a secret, I think."

Tanaka's reluctance was understandable. High-temperature superconductivity has enormous commercial, as well as scientific, promise, and the leaders in the field will have the keys to a treasure chest of applications. So far, the only commercially viable superconductors are cooled with liquid helium, which is both expensive and hard to handle. Because the new materials can transmit electricity without loss at higher liquid-nitrogen temperatures, they have numerous potential advantages over the old-style superconductors.

"The race to commercialize superconductivity is on," U.S. Energy Secretary John S. Herrington told a 1987 federal conference on commercial applications of the new science (not incidentally, foreigners were barred from attending). "And the economic prizes await the nation which first discovers a viable, marketable technology. We will face unprecedented international competition. Although we have the jump on our competitors in basic research, we must marshal all our resources and tap our ingenuity to the fullest to compete effectively in the marketplace. Winning medals and prizes for our scientific research is some consolation, but it isn't enough."

But the distance between the laboratory and the marketplace is often long. Space engineers didn't invent rockets one day and send men to the moon the next. Some twenty years elapsed between the discovery of the double-helix structure of DNA and the first transplantation of a gene from one organism to another. The atomic bomb came forty years after Albert Einstein gave us the theory behind it and his formula that linked energy and matter: $E = mc^2$. As Donna Fitzpatrick of the Department of Energy recalls:

> When the transistor was invented, everybody said, well, that's kind of interesting, and so it was licensed to the Japanese because we didn't know exactly what to do with it. Now just about every home appliance has a transistor in it, and a little microprocessor to program it, your microwave oven, VCR, and a lot of other things. Some people have said we probably haven't even thought of the applications that will have an impact. Who

knew, thirty-five, forty years ago, what the transistor, the integrated circuit, the semiconductor, would lead to? My recurrent horror story is that the Japanese will market a superconducting wristwatch. It won't really be superconducting, but it'll have this little chip in it, like a digital watch, a little microprocessor chip, and some little corner of it will have some new superconducting material in it, and sure, at some point in the circuit the electron will run through a bit of superconductor, and we'll buy it from them.

But applications, including that hypothetical wristwatch, are a ways off. Scientists still don't know how the new superconductivity works. The materials are still very brittle, and until they can be bent, twisted, stretched, and machined—and made to carry large amounts of current—they will be as valuable to industry as a gold brick to a sailor on a desert island. Lousy, in fact, is the way metallurgist Masaki Suenaga of the Brookhaven lab has described the new materials, including one with a T_c between 90° and 100° K that he and an associate discovered. "The new materials are deceptive," he said. "Because they are easy to make and will levitate a magnet in liquid nitrogen, they give the impression that it will be simple to turn them into practical use. That impression is wrong."

Suenaga may even be understating the problem. Turning these materials into *any* use, practical or otherwise, is a job and a half. It will be hard enough just to shape and fire the ceramic superconductors into the right form—wire and tape that could be wound into cables to transmit electricity and into coils to provide large magnetic fields, thin films for computer chips, and an array of tubes, pellets, cylinders, even bowls. At the same time, the ceramists have to line up the crystalline structures in these complex substances in just the right way if they are to transport enough current to be of practical use. The materials are, after all, chemical club sandwiches, with layers of copper and oxygen separated by barium, lanthanum, yttrium, and all the other fillings that the ceramist-chefs can concoct. "It's not just one sheet of material you're talking about here," the Naval Research Lab's Don Gubser said. "It's building up an oxide layer, then another sheet, then another. It all becomes yet another integrated circuit package."

While no one is saying it will easy, researchers are somewhat consoled by the fact that one of the common, old superconductors—niobium-tin, an extremely brittle alloy—is wound into thousands of miles of wire every year. And so in laboratories throughout the world, ceramists grind their new materials into fine powders, fire them black in hot ovens, then blend them with binders, plasticizers, and wetting agents (a popular binder is oil of menhadden, a herringlike fish used as bait and fertilizer).

What this blending technique really boils down to is embedding the ceramic's superconducting microscopic grains in a plastic base. The resulting glop looks, in its early stage, like thick black paint, and later, when it is dried, like lumps of Silly Putty or Play•Doh, begging to be squeezed into some shape or other. At Argonne, the lab in which they do all of this looks like a paint store, a machine shop, and an old-fashioned pharmacy all under one roof. Plastic bottles of black goo rotate slowly on mixing rollers. Beakers of solvents and polyethylene glycol and jars of earthy materials line metal shelves. There are dirty mortars and pestles, an array of dies and bench presses, assorted wrenches and tongs, hammers, mallets, and screwdrivers.

A technician demonstrates how tape is created from some of the plastic batter in the mixing bottles. He pours a batch of the goo onto a glass plate and carefully runs a metal spreader over it, flattening it into a thin sheet that looks like latex-based paint. It dries, and he peels it off neatly. It is flexible, waxlike to the touch, about .002 inch thick.

At a bench press, another technician loads a lump of the same material—but dried to the consistency of putty—into a standard laboratory press, forcing it through a vermicelli-shaped die. Out comes "wire," about twice the thickness of a human hair.

The technician is accustomed to visitors watching him work ("Hey, I saw you on TV in Japan," a Japanese journalist at the demonstration comments). He dismisses his technique with a shrug. "All you do is put that goop in there, push that, and there you have it. Someone on one of those TV science shows called it a biotech feedback loop, whatever that is."

The thin strands look and feel like the black thread cobblers use to stitch soles, but they are still flexible, or "green," because they haven't yet been fired. When that's done, the material is wire of a sort, but it soon loses its flexibility—as does the tape—and becomes brittle enough to snap easily. It is quite obvious, twisting a bit of it between two fingers, how difficult it is going to be to wind it around some giant magnet or stretch it into something capable of carrying enough electrical power to light a city.

"We've improved the material's strength by using a finer powder, reducing the binder content, and controlling the firing schedule more carefully," explains Roger Poeppel, who manages Argonne's division of materials and components technology. "These changes have decreased the ceramic's brittleness by at least 50 percent in a few months, but it's still a long way from being truly flexible. There's really no forgiveness at all in this stuff. It's not like a piece of copper wire that you get a kink in if you screw up." At this writing, Argonne ceramists have only been able to make three-foot lengths of ceramic wire, but they have wound it into coils and fired the whole chunk, a technique that could one day provide the very coils required to produce magnetic fields large enough to be used in, say, nuclear magnetic resonance (NMR) equipment, the helium-cooled scanners that doctors use to diagnose disease.

In an attempt to make wire better suited to commercial application, ceramists and metallurgists at Argonne and elsewhere are working hard to develop composite wires, bonding the ceramic to an outer coating of, say, a nonsuperconducting metal, like copper. If the ceramic were to lose its superconductivity and become resistive, the copper braided through it might take over and continue to carry the current until the superconductor bounced back.

Another novel approach has been tried at Ohio State University and at Cambridge University in England. Both groups use silver, which has the lowest electrical resistance at room temperature; as a conductor of heat and electricity, it is far superior to all other metals. At Ohio State, physicist James Garland and a graduate student, Joseph Calabrese, add powdered silver to their ceramic mix and, by heating it, manage to fuse the silver particles into a solid structure that is randomly distributed throughout the ceramic, sort of like a series of tiny bones in and around which threads of superconducting material flow.

The metal-ceramic composite appears to have a number of advantages: it improves the mechanical properties of the material, allowing it to retain more of its superconductivity while making it more plastic; the silver matrix protects the composite from the degrading processes that result from exposure of the material to moist air; the metallic silver allows the material to cool quicker and more efficiently than the composite alone.

There's also the possibility that the silver mixture will allow the "superconducting proximity effect" to occur. (Superconductivity can occur between two superconductors that are physically separated so long as the barrier is a conducting metal.) That tunneling effect may be the key to developing materials that can be machined, drawn into wires, or cast into various production shapes. Preliminary experiments have shown that the new silver composite is quite strong and can be machined into several useful forms.

At Cambridge's department of physics, Jan Evetts—who has filed at least six patents ranging from the basic physics of ceramic material to designs for superconducting wires— packs ceramic into narrow tubes of silver to create flexible, superconducting wire. Because the silver doesn't oxidize, it allows the oxygen to pass through to the ceramic, which requires it.

Evetts also tells a story that underlines the highly competitive nature of the ceramic wire business. Within days of the filing of one of his patents, a Japanese trading company contacted him and offered research funds in exchange for patent rights. A few English companies also expressed interest, but their approach was different. According to Evetts, "The English said, 'Can we have lunch?' while the Japanese said, 'We want to give you money.' "

Far more dramatic than stuffing ceramic into tubes, however, is a technique that uses relatively inexpensive explosive technology to make superconducting sandwiches of metal and ceramic oxides. It's a concept known variously in the trade as dynamic compaction, shock compression science, or more simply, explosive welding. Layers of copper or aluminum serve as the "bread" for the sandwiches, with powdered ceramic as the filling. Shock waves from an ammonium-nitrate explosive (which costs twenty-three cents a pound) bond the materials into superconducting "monoliths," long, metallic masses that can be joined together in any length desired. This type of arrangement, called a transmission bus, would be placed inside an insulating tube with liquid nitrogen as the coolant, then buried to serve as a current-carrying transmission line.

According to Lawrence Murr, a materials science professor at the Oregon Graduate Center, the use of shock compression is a concept in search of applications, and the new oxides are especially amenable to such an unconventional manufacturing method. "You start with any cheap oxide," he said. "The only labor-intensive thing about the process is putting the design together. You can take the manufacturing process and design the shape you want very easily. If we understand the application, then we can match the manufacturing process with it."

Some scientists have foregone the ceramics altogether and attempted to fabricate wire out of metal alloys that can superconduct at temperatures warmer than normal for a metal. At the Massachusetts Institute of Technology, for example, one team has made a superconductor from europium—a soft, silvery metal that was once in short supply but is now more easily acquired—barium, and copper. The alloy apparently works well at 90° K, which means it can run on liquid nitrogen; moreover, it is more easily fabricated than the ceramics and could be spun into ribbon from a liquid form, oxidized, and turned into a superconductor.

But wire, monoliths, cables, and coils are not the only shapes of interest to materials scientists trying to shape the new superconductors into some useful form. Thin films, the form that the ceramic oxides would take in advanced computers and a host of electronic devices, are actually further along than the ceramic wire and indeed may be the very first of the new superconductors to see a practical application.

Thin, however, is an understatement. To the researchers who make the superconducting films, *thin* means a micron, which translates down to a few thousand atoms thick. And that's for the complete film. Each of the layers that make up the film must be only a few angstroms thick, which means but a *few* atoms. Getting to that level is extremely difficult, but one team from Bellcore and Rutgers University has succeeded, though it took them five weeks of sixteen-hour days to do it.

Working with the yttrium mixture, the researchers placed it in a vacuum and fired several thousand laser shots at it, ten pulses per second. The process, called pulsed excimer laser evaporation, produces a vapor from the superconducting compound with each shot. It is the vapor that is deposited on a sample material to form the ultrathin layers. After the film has been built up, it is baked at high temperature, and when cooled, it shows a large reduction in electrical resistivity beginning at about 90° K (the so-called onset temperature) and full superconductivity (zero resistivity) at 83° K. These temperatures are in the same range as in the original bulk material.

According to Venky Venkatesan, Bellcore's research manager for the project, the new process can work with any yet-to-be-developed superconducting materials. "It's a very basic process," he said. "You name the bulk material, we can shoot at it and make a thin film out of it. And more important, it will have the same ratio of elements that it did in bulk form. Using a laser preserves this ratio, while other processes don't."

The laser method is also the least expensive method yet developed for making thin superconducting films. Cost savings, then, may become more substantial later on, said Venkatesan, because the pulsed method lends itself to mass production. "The lasers we've been using are very low power, compared to what else is out there. With larger, more powerful lasers, thin films can be deposited over larger areas in even shorter amounts of time."

Other researchers, notably at IBM, have relied on techniques that spray-coat the superconducting ceramics directly on almost any kind of backing material, including silicon chips. This method, known as plasma spraying, involves ionizing a jet of powdered oxides in an electric arc just over the surface of the backing material. When the plasma condenses, it leaves a strip a few centimeters wide and a few micrometers deep. By spraying through a template, very fine superconducting "wires" can be deposited on a chip, turning it into a circuit board that could serve as a link between chips in a computer cooled in liquid nitrogen.

Such high-tech methods have not yet brought the new superconductors into the marketplace, but they most certainly give manufacturers what they need most from the warmer bulk ceramic materials—the forms that will carry frictionless electrical current and channel its perpetual, enormous power into everything from that wristwatch to flying trains. If the necessary properties can be built into the new superconductors, if the problems that still bedevil them are ironed out—and few, if any, believe that they will not be resolved—then superconductivity, once an exotic plaything, will, like the transistor and the laser, change the very way we live and work.

Dr. Praveen Chaudhari, vice president of science for IBM's research division, addressing the U.S. House of Representatives' Science, Space, and Technology Committee in the spring of 1987, said:

> Certainly, there is no lack of enthusiasm on the
> part of scientists throughout the world to under-

stand these materials and overcome the remaining problems. And it is equally certain that every scientific breakthrough will be fully explored by the engineering community, as it strives to develop practical uses for the technology.

I've heard many speeches and read many predictions of levitating trains and cheap energy. It is difficult, if not impossible, to contain the excitement generated by all the recent advances. Yet, we must all remember that it is a long way from a small magnet levitating over a Petri dish to a large train levitating over its tracks.

The possibilities, however, are indeed as awesome and wide-reaching as the imagination can absorb, and they may affect the lives of all of us one day. In my twenty-one years as a scientist, I've never experienced anything as profound.

6

CURRENT PROBLEMS

SOMEDAY, THE NEW "warm" superconductors will be used to replace conventional conductors in anything electric or electronic—computers, transmission lines, the magnets in medical imaging devices, generators in a car, a factory, spaceship, or a hydroelectric dam. "All kinds of switches and sensors," is the way Donna Fitzpatrick of the Department of Energy puts it.

> Anything that's useful in computers will get into telecommunications because they're virtually the same thing. Anything which involves sending and receiving a signal is going to become an area of application because a superconductor is minutely sensitive to any kind of electrical current. It's like if there were a brick wall between us, and I blew on my side of the wall, you wouldn't know it because the wall has so much resistance to my breath. But if there's a sheet of cellophane, say, between us, and I blow on it, you'll know it because there's no resistance. So in a supercon-

ductor, if it has an electron coming along ever so
gently, the electron's going to get through where
it wouldn't get through in an ordinary conductor.

It does sound wonderful, and there is little doubt those
unimpeded electrons will one day be tapped and channeled
into doing some very useful work. But such success will
depend on a number of factors, and some of them, for the
moment, pose nagging problems.

Achieving the right arrangement of oxygen in the lattice
of the superconducting ceramics, the necessity of converting the crystals from tetragonal to orthorhombic symme-
try, the relative fragility of the ceramic wires, and the lack
of a solid theory to explain how they work are all, as we
have seen, important considerations.

But there is something else that troubles the materials
specialists trying to fashion the ceramics into something
useful. It's a question that even nonscientists can easily
understand: how much juice can they carry? For without
substantial current flowing through their labyrinthine
ways of chains, planes, and grains, the new superconductors are worthless. Academically interesting as an electrical phenomenon, yes, but like veins without blood. And the
difficulty at the moment with the new materials is that
they cannot yet carry enough electrical current to make the
revolution that everyone is predicting. "These things are
about as conductive as a piece of balsa wood," one Brookhaven scientist told me in frustration. "Only don't quote
me."

The ceramics are not, of course, as bad as all that. They
most certainly do superconduct. But just barely. Subtle
shifts in oxygen balance, too much or too little of one
ingredient or another, very high magnetic fields, and high
temperatures—all of these reduce their unique capabilities
and turn them quickly into insulators or, at best, conductors that offer no real advantage over those spun of more
reliable copper wire.

The attempt to boost their current-carrying capacities is
on yet another fast track, one that runs parallel to the road

to room-temperature superconductors. While no one is ecstatic over the results thus far, enough signs indicate that vast improvement in the performance of the new superconductors is inevitable.

But there is a key, interrelated variable that, along with temperature, influences superconductivity: magnetism. Immediately the word summons up an image of a horseshoe-shaped piece of metal that almost magically pulls in iron filings and other metallic objects. Or perhaps a towering railroad crane lifting scrap metal from a jumbled heap and swinging it over an empty flatcar. Or a compass, whose quivering needle always lines up in the same direction no matter which way its case is turned. Albert Einstein, intrigued by a compass when he was a child, would write later in life that its steadfast needle made him aware for the first time that "something deeply hidden had to be behind things."

Long before Einstein, however, perhaps before written history, magnets and magnetism were stirring the human imagination. The ancient Greeks, certainly, knew of the power of lodestone—the hard, black iron oxide, magnetite—to attract iron, though they didn't know why or what to call the phenomenon. It was Pliny the Elder, the Roman author and naturalist, who tied the word *magnetism* to a shepherd named Magnes, who lived in Magnesia, an ancient land in Asia Minor with a plentiful supply of magnetite. As the story goes, Magnes was walking through a rock-strewn field and was astonished to find that bits of rock were sticking to the iron nails in his shoes and to his iron crook.

But while its semantic origins fit well into a game of trivia, the attempts over the centuries to understand magnetism's power have been no trivial pursuit. Three thousand years ago, the Chinese discovered that a piece of lodestone, hung on a string, would always line up in a north-south position and could be used to navigate a ship. In A.D. 1269, a French scientist, Petrus Peregrinus de Maricourt, carried that further, tracing the lines of force on the lodestone to discover its two poles. Compasses worked, it turned out, because the earth is a gigantic magnet with its own magnetic field, a discovery made in 1600 by William Gilbert, personal physician to Queen Elizabeth I.

As important as these discoveries were, however, fuller understanding of magnetism came only after electricity had lost some of its mystery and after scientists stopped treating the two phenomena as separate entities. Over time, they would combine them into a new branch of physics, electromagnetism, whose prime tenets were these: an electric current produces a magnetic field, and conversely, a changing magnetic field can induce electric currents. In 1820, the Danish physicist Hans Christian Oersted was the first to establish the truth of the first with a simple experiment. By holding a wire parallel to a compass needle and passing a current through the wire, he deflected the needle a quarter turn, so that it came to rest at right angles to the wire. Electricity did create a magnetic field. The second principle, that magnetism produces an electric current, was demonstrated independently ten years later by physicists Joseph Henry in America and Michael Faraday in England. Simply by moving a magnet near a coil of wire, they were able to produce a fleeting electric current by means of what we now know as electromagnetic induction—the creation of current within a conductor by a changing magnetic field. So long as the magnet was kept in motion near the wire, current was produced; when the movement of the magnet was stopped, current stopped.

This discovery that electricity could be "made" without a voltaic battery or the use of any chemicals at all would set the stage for the development of one of the world's most notable power-generating machines, the electric generator, or dynamo. Mechanical energy, from a steam engine, say, or from water power, spins a coil of wire inside a magnetic field, generating electrical current. All generators can be run in reverse as electric motors: Current sent through a coil wire in the generator creates a magnetic field in a coil of wire known as an armature winding; other magnets around the armature continuously attract and repel it, and the armature spins. A shaft connected to the spinning armature is then able to deliver the mechanical energy that the motor has produced from electricity.

Those are some of the benefits of controlled magnetism. But magnetism does not always work to our advantage. Medieval legend tells of the magnetic mountain which drew out all the nails of any ship that neared, sending the vessels to the bottom. More scientific, however, are the disturbances called magnetic storms, occasional disruptions of the earth's magnetic field correlated with sunspot activity. These effects, though, are temporary, and the earth's field manages to recover after a few days.

Far more serious—and far more consistent—is what magnetism can do to superconductivity. It is, in fact, its bane. At least it is in the old superconducting materials.

We've said that because a superconductor offers no resistance to electrical current, it does not require enormous amounts of power to sustain the flow of current. Power is needed only to establish the flow. Because they do not dissipate electricity, superconducting materials are perfect for making the windings that go into the coils of high-field magnets. But there is a catch-22. When a metal becomes superconducting, it produces, as do ordinary conductors, a magnetic field. And whenever a current strong enough to stimulate a reasonably high magnetic field is sent speeding through the material, the field that is generated seeps inside and destroys the superconducting effect. Even weaker magnetic fields can quench the benefits of the superconductor, and the point at which this occurs is called the critical magnetic field.

The degree to which a superconductor can be ruined by a magnetic field depends on the type of conductor. There are two: Type I superconductors, of pure metallic composition, and Type II, mostly alloys and, now, the new oxides. Generally speaking, the Type I superconductors have a lower critical magnetic field—that is, they lose superconductivity at relatively lower field strengths—than the Type II superconductors. This is a vital point, because if a superconductor is to be wound into coils for a large magnet, the wire has to be able to withstand an intense magnetic field of more than 2 tesla, which is forty thousand times stronger than the earth's magnetic field. (The measurements of magnetism that are generally used are tesla and gauss. A tesla is equal to 10,000 gauss. By way of reference the following values might be helpful: the earth's natural magnetic field, the force that pulls the needle of a compass, is a paltry 0.5 gauss, or 0.00005 tesla; the common magnet that pins notes to the refrigerator door has a field of a few hundred gauss. The present-day, widely used superconducting magnets wound with niobium-titanium are able to sustain a field of 12 tesla, or 120,000 gauss and higher—an enormous level, considering that conventional electromagnets cannot operate much above 2 tesla.)

Happily, the new superconducting materials have extremely high magnetic fields—meaning that it takes a field of stupendous intensity to completely restore electrical resistance. Precise measurement of such large fields, however, takes some doing, since the fields of present research magnets, high as they are, are not limitless, and estimates are about all scientists can rely upon.

One place—indeed, it is *the* place—where the critical field of the new oxides comes under intense scrutiny is the Francis Bitter National Magnet Laboratory in Cambridge, Massachusetts. Operated by the Massachusetts Institute of Technology in a huge old brick building that was once a bakery, the lab was established in 1960 as the first center both for conducting research on high magnetic fields themselves and for providing the high fields that have become essential tools of the experimental physicists.

Today, it is a mecca for visiting scientists bent on learning more about the complexities of magnetism, and eager to expose their materials to the massive fields unavailable elsewhere. Teams of physicists and materials specialists— they range from NASA types investigating changes in magnetic properties of space-lofted materials, to the occasional inventor building a better can opener—descend regularly on the lab. They confine themselves for a week or so to cell-like stations that are plugged into twenty-five water-cooled magnets with high-field strengths up to 23 tesla, a pull more than five hundred thousand times stronger than the earth's. A hybrid magnet, part copper, part super-conducting niobium-titanium, can achieve a towering 30 tesla. There's also a helium-cooled, 12-tesla, full-supercon-ducting magnet that the lab bought to relieve some of the scheduling load from the lab's Bitter magnets. It requires no power input ("It just sits there and works for twenty-four hours," said a technician).

What makes the Bitter magnets so special (they're named after the late Francis Bitter of MIT who conceived them in the late 1930s) is their design. Standard iron-core magnets are just large yokes of soft iron with coils wrapped around them. They produce their magnetic fields by magnetizing the iron, which is easy to do, thanks to iron's properties. However, at about 3 tesla, iron "saturates," and the field cannot go much beyond. So Bitter got rid of the iron. His design stacked layers of pierced copper disks, with insula-tion (Teflon and fiberglass are used today) between them, a structure that allowed large amounts of water to pump through the magnet and wash away the heat quickly. Next, Bitter got his hands on a surplus trolley generator, placed it on the MIT campus, and hooked it to his perforated stack. It gave him a steady field of 10 tesla, the strongest continu-ous field achieved at that time. The hybrid magnet men-tioned above is a 21-tesla Bitter-type copper magnet wrapped in a 9-tesla superconducting magnet made of niobium-titanium. The 30-tesla field at the magnet's core represents the sum of all the fields generated by the coils around the core, as is true in all magnets.

In the lab, the floors of the cells that house the Bitter magnets are strewn with octopuslike tentacles. These thick black hoses suck in 1,000 gallons of water a minute from the nearby Charles River to cool the magnets, and then discharge it back into the river, where it winds up a luke-warm current. Power cables linked to four throbbing gener-ators with 80-ton flywheels snake in and out, carrying 10,000 amps per cable, sending as much electricity into a 230,000-gauss magnet as that generated by a conventional power station. Twisted vacuum lines feed the stations with the supercold liquid helium required for the experiments. Wisps of white vapor hover eerily in the air around the stations as technicians sporadically check liquid helium levels in the sealed Dewar jars that prevent the helium from following its natural inclination to become a gas again. On one wall hang shining copper disks, some of them blackened by the intense forces they've been subjected to in the cores of the powerful magnets. The noise from the generators, the cleansing machinery (the Charles's murky water, though plentiful, is not fit to send through a magnet), the pumps, the heat exchangers, and the apparatus that reliquefies boiled-off helium is constant and deafening a discordant chorus of roaring, throbbing, clanking, squeaking, whooshing, hissing, whirring, ticking. It is difficult to understand how anyone but the hearing-impaired can remain in here for very long, let alone think esoteric thoughts.

"It doesn't take all that long, really," physicist Donald Stevenson shouts over the din. (A sign outside warns that the noise is at a level which requires ear cover if the experimenter is inside for more than 90 minutes) Pointing to the small opening at the top of the gleaming magnet stack, he explains:

> This is the business end, a 2-inch diameter bore, helium around it. Those hoses carry purified water in to get the heat out. Over here are water-cooled cables that supply the current. Halfway down the structure, the maximum field occurs. The sample goes in here, through the bore hole. Typically, these guys bring in an experiment, say a sample of a superconductor they've made, to see how it will behave in a magnetic field. They stick it in. A half hour can do it. If they're experimenting with all kinds of compositions, they might need a whole day to test, say, fifty samples. To get a continuous field as high as 30 tesla, we go to 1-second pulses, which doesn't require a huge amount of power, and this gives us a field for one, two, three hours.

Our highest field now is 30 tesla, and we can go to stronger fields—but only if the experimenter is willing to have it on for a fraction of a second. The problems with strong magnetic fields are, first, the heat that's given off. You've got to take that away, and that's why all the money we spend to buy power from the Cambridge Electric Company, a half-million dollars a year, winds up as just a little tepid water in the Charles. Thermodynamically, magnets are zero efficient, you know. You put a lot of power in—all that power in that generator room ends up in this—and you get no useful work out, just this heat. The other problem with a strong magnetic field is the strong forces generated. The stronger the field, the more the danger that the magnet will blow apart. So if you're willing to go for a 10-millisecond test with your sample, you can, because the magnet doesn't have a chance then to heat up and melt. You've got to be fast.

It was in these Bitter magnets that the ceramic superconductors showed their magnetic mettle. Early tests indicated that the critical field was roughly proportional to the critical temperature, which meant that the new materials could be expected to have critical fields four to five times higher than traditional superconductors. From a practical standpoint, this in turn meant that much stronger electromagnets could be built, and superconducting power transmission lines could carry much higher currents.

Subsequent testing confirmed that the new oxides were, indeed, quite able to handle very high magnetic fields. In one assessment by Bell Lab scientists, the critical field of a barium-yttrium-europium–copper oxide ceramic was shown to be 50 tesla (500,000 gauss) at 77° K, a whopping million times the earth's magnetic field! By way of comparison, the highest critical field among commercial superconductors wound from niobium-tin is 35 tesla (350,000 gauss), and that's near absolute zero. For niobium-titanium, superconductivity can be destroyed at around

120,000 gauss. Other estimates for the new materials range even higher—anywhere from 800,000 gauss to better than 3,000,000.

But just as magnetic field strength affects superconductivity, so too does excessive electrical current. If current is passed through a superconductor and increased, at a certain point superconductivity is destroyed. Thus, the major limiting factor for commercial applications of the warm superconductors is what we have referred to as critical current density—the measure of how much current the material can carry before losing superconductivity. And the critical current densities for the new oxides are disappointingly small. The materials simply don't carry enough electricity to be useful yet on a large scale. As noted earlier, to make a practical material that would operate in the liquid nitrogen range, the current density must be increased at least one hundred times.

Just as the critical temperature and critical magnetic field are rising steadily, so too is current density increasing. It is true that most applications require an operating current density greater than 10,000 amperes per square centimeter, which is ten to one hundred times higher than densities measured thus far in ceramic wires and tapes. (The filament of an electric light bulb carries 1,000 amps per square centimeter.) But single crystals and thin films have far surpassed that—they've shown current densities greater than 1,000,000 amperes per square centimeter, which puts them in the same ballpark with the top range of today's applications in magnets and some electronic devices. Bulk samples of some of the newer materials have also been forced to carry more current by improving the way they are processed.

The ability to grow single-crystal samples of the new materials is regarded as an important milestone in current research, enabling researchers to fine-tune their observations of the ceramics' current-carrying properties without the complications caused by poor electrical connections between individual grains in a polycrystalline sample. Single-crystal bits like these will also play a big part in making a variety of new microelectronics and other devices.

Argonne was the first lab to report the growth, crystal structure, and superconducting transition temperature of a single crystal of the yttrium compounds. The first ones were only 0.3 millimeters long, but they packed a wallop.

As Merwyn B. Brodsky, acting director of the Materials Science Division at Argonne, put it: "Our single crystals had current densities of more than a million amps per square centimeter, much larger than had ever been seen in a polycrystalline material." But there was also discouraging news. When the temperature around the crystal was raised to 77° K, nitrogen's boiling point and the range targeted for most applications, current density fell significantly. Moreover, the superconducting current seemed to flow better in a particular direction. Studies at the National Bureau of Standards (NBS) indicated the same thing. According to Jack Ekin of the Electromagnetic Technology Division, the extreme sensitivity of the new superconductors to magnetic fields, as well as some other shortcomings, might be due to the disordered arrangement of crystallites, or grains, in the materials. NBS's early data suggested strongly that the materials contained strong and weak superconducting, and nonsuperconducting, regions. The poor areas of superconductivity seemed to occur at the boundaries between the crystals making up the materials.

Like a chain, then, the ceramics appeared to be only as strong as their links. If the strong regions were arranged in a sequence of planes, the superconducting current would flow unhindered from grain to grain. If forced to jump to a plane stretched out above or below, as with a perpendicularly arranged crystallite, the current would drop. Thus, the geometric arrangement of crystals was crucial if current density was ever going to be increased.

IBM researchers managed that trick in 1987, reporting that they had attained a critical current of over 100,000 amps per square centimeter in a thin film of one of the yttrium mixtures cooled by liquid nitrogen. The experiment was accomplished by growing single crystal samples and large grains of the materials in which the crucial atomic planes were oriented in a single direction, the one which best accommodated the flow of superconducting electron pairs. Japanese researchers working with the same material at Nippon Telephone and Telegraph eventually topped that—by a factor of eighteen, to 1,800,000 amps per square centimeter.

In January 1988, Sumitomo Electric Industries, Ltd., announced they had made a superconducting ceramic film of holmium—a soft metallic element with no previously known uses—that attained a maximum current density of 2,540,000 amperes per square centimeter at 77° K. According to the Japanese, the new film maintained its superconducting properties a month after fabrication and could pass a strong current even while under the influence of a magnetic field measuring 10,000 gauss. Moreover, they said, the ceramic was able to achieve its high current density because it allowed electricity to pass through it both vertically and horizontally.

Significantly, the best results thus far have all been achieved with thin films. The goal, of course, is to arrange crystals together in the right order in bulk material, an intricate and expensive process that has not yet been accomplished. "You've just got to distinguish between atomically coherent single crystals, which carry respectable amounts of current, and bulk superconductors, which are really compressed jumbles of separate crystals," said Alan Wolsky, a senior energy systems scientist at Argonne. "We would dearly love to get the bulk up to those single crystals."

Since superconducting current flows predominantly along a single plane in a crystal, many labs are trying various "texturing" techniques to arrange crystals in superconducting planes; the resulting polycrystalline sample would, it is hoped, mimic the behavior of a single crystal. Some methods train magnetic fields on the crystals as they form, the idea being to nudge them into rows; others spread thin layers of the ceramics on alloy-templates tailored to the right crystal alignment. Argonne scientists have prepared bulk materials with some texturing and achieved improvements in current density from 10 to 30 amps per square centimeter at 77° K, in a magnetic field of around 4 tesla, some eighty thousand times as strong as the earth's.

"These improvements are small," said Merwyn Brodsky, "but they show we are moving in the right direction. They suggest that critical current can be improved by increased texturing. Critical current drops sharply when small magnetic fields are applied, but after this initial drop, current density remains extremely stable in the textured materials as magnetic field is raised to higher values."

Another barrier to current flow is the movement of fluxoids, magnetic field lines that penetrate the superconductor, concentrate themselves in various regions, and interrupt the orderly flow of current. Scientists have already devised techniques to immobilize the fluxoids, allowing the rest of the material to superconduct. Here, too, Argonne researchers managed some successes by focusing on defects in which fluxoids appear to be captured. (Examples of defects are empty oxygen sites in the ceramics, and "twinning defects," which occur at boundaries of crystals that are mirror images of one another.)

In one experiment, they annealed the material in an oxygen atmosphere, allowing atoms to rearrange themselves naturally and "heal" any defects that might capture—or to use the physicist's term, pin—magnetic flux. Because annealing led to a marked decrease in critical current, this suggested that flux pinning at defects played a significant role in determining critical current. Pressing their experiments further, the researchers deliberately introduced metallurgical defects. The results confirmed their theory: the artificially created defects paralyzed the magnetic field regions, thereby enhancing critical current two to five times. The biggest improvement occurred, moreover, at 77° K, and in the direction perpendicular to the copperoxygen planes. "Taken together," said Brodsky, "these two approaches show that critical current can be controlled by regulating the material's defect content, and that substantial improvements can be expected if proper defects can be introduced."

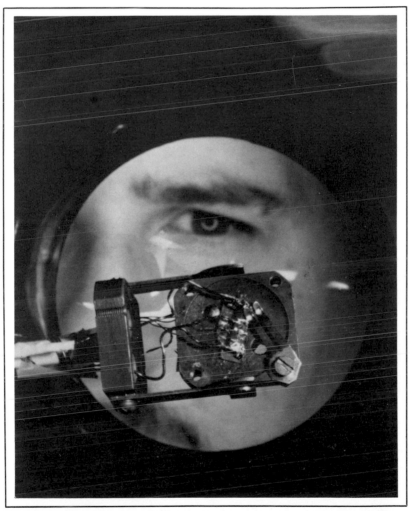

Scientist at Argonne National Laboratory examines a test cartridge holding a high-temperature superconductor. (*Photo courtesy of Argonne National Laboratory*)

THE BREAKTHROUGH TEAM. IBM's Georg Bednorz and Alex Müller in their Zurich Laboratory. (*Photo courtesy IBM Corporation Research Division.*)

A batch of ceramic powder before it is molded into superconducting wire. (*Photo courtesy of Argonne National Laboratory*)

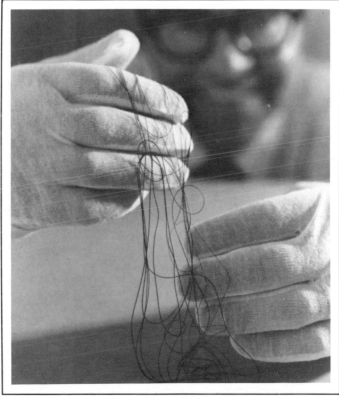

SUPERWIRE. This superconducting wire, made from a ceramic material, is still flexible because it has not yet been fired—a process that makes the new materials brittle. (*Photo courtesy of Argonne National Laboratory*)

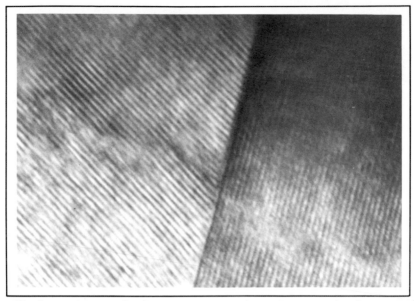

Electron micrograph of copper-oxygen planes in a sample of super-conducting yttrium-barium-copper-oxide. Planes are well-defined up to the grain boundary. (*Photo courtesy of Argonne National Laboratory.*)

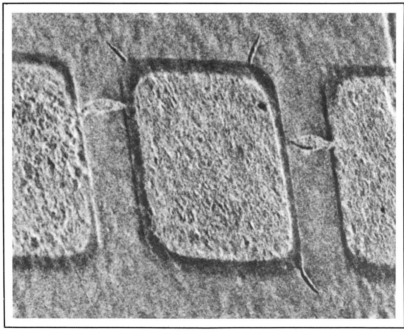

Here, magnified more than 500 times, is a high-temperature superconducting thin-film electronic device known as a SQUID. It can be used for extremely sensitive magnetic measurements and is only one one-hundredth the thickness of a human hair. (*Photo courtesy the IBM Corporation Research Division.*)

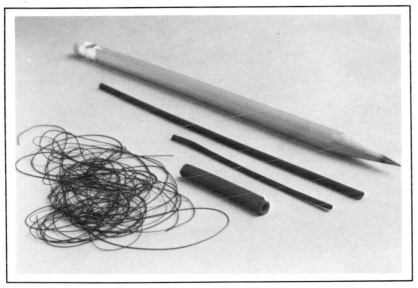

Objects made of high-temperature superconducting ceramics include wires, cylinders, pellets, and tubes. (*Photo courtesy of Argonne National Laboratory.*)

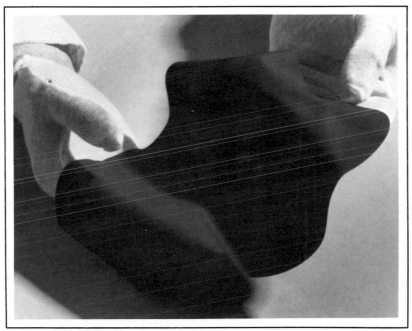

A tape of high-temperature superconducting ceramic. (*Photo courtesy of Argonne National Laboratory.*)

Superconductivity is demonstrated here as a superconducting ceramic pellet floats above an array of magnets. (*Photo courtesy of the Argonne National Laboratory.*)

History being made at Argonne as a scientist prepares to send electrical current through a superconducting wire at −321° F (77° Kelvin) for the first time. (*Photo courtesy of Argonne National Laboratory.*)

Meissner Motor developed at Argonne uses yttrium-barium-copper-oxide to run. The 8.5-inch aluminum plate has 24 small electromagnets mounted along the outer edge and two hockey-puck-shaped disks made of the ceramic. (*Photo courtesy of Argonne National Laboratory.*)

A test model of the Japanese Magtrain shown here at a world science fair in Tsukuba, Japan.

The Navy's *Jupiter II* on a shakedown cruise. More than a dozen at-sea trials have demonstrated the feasibility of superconductive electric propulsion. (*Photo courtesy of the U.S. Navy's David Taylor Research Center.*)

7

CHAINS, PLANES, AND GRAINS

IT IS SURPRISINGLY easy to make a superconductor out of one of the new ceramic materials. Don Gubser of the Naval Research Laboratory says somewhat facetiously:

> You simply go down to your local chemist's supply store and buy some copper oxide, some barium carbonate, and some yttrium oxide, all readily available in white powders. You go home and you do a little calculation so you'll get the ratios right, so that the yttrium, barium, and copper are in a ratio of 1-2-3. If you've had any chemistry at all, that's very easy to do. Then you simply stir the mixture thoroughly in a pan and stick it in your microwave oven and you bake it at 950° F for about four hours—and when you pull it out, you've got this black powder. You compress it into a little plate, and you put it back in the oven to sinter it and make it one solid piece. And if you know a little more about what you're doing, you blow some oxygen on it and

cool it all slowly—in the old days, like a few months ago, we'd just stick it back in the oven, let it sit, and just bring it back out.

Next you drop that little plate into liquid nitrogen. You can get that in any small town in the country, it's that plentiful. In fact, I was giving a talk in Iowa one day and I said I couldn't finish the demonstration without liquid nitrogen and I couldn't bring any with me. So someone ran out to the vet store and got some. The store had been using it to store bull sperm for artificial insemination.

So, you drop that little disc of ceramic into the liquid nitrogen—you can put it all in a styrofoam cup—and put a little magnet on top. If the magnet floats, you've got a superconductor, the Meissner effect.

That casual bit of instruction is fine as far as it goes. But making a high-quality ceramic superconductor, one that has a high critical temperature and a lot of superconducting regions, takes a lot more work, and you truly do have to know what you are up to. Standard laboratory instructions are replete with warnings about the toxicity of barium carbonate ("Take appropriate precautions, wear a mask, avoid contact or ingestion, check with a chemist for safe procedure"), and they're full of exacting directions about how to press the material into a pellet at 10,000 psi with a die and hydraulic press; how to sinter it inside a quartz tube so that it has only a few points of contact with the quartz; how to encapsulate the stuff with varnish or epoxy to protect it from moisture (when the yttrium and lanthanum mixes, especially, react with water and carbon dioxide, they easily lose their superconductivity); and how to make the electrical contacts with alligator clips or indium-soldered copper leads. Not something, obviously, that one would want to bother with if the only goal is to perform a fairly simple lab demonstration.

Fabricating a really good superconductor hinges on so

many variables. One is the intricacies of the material's microstructure, a configuration made even more complex the more ingredients—thus more atoms—there are. Even when the positions and interrelations of all the atoms seem to be correct, not all samples are alike, and they may not even superconduct, even though the formula says they will. Or they may superconduct even when the molecular structure indicates they should not—something that has been noticed in some of the new superconductors. Another consideration is that the oxygen content has to be carefully controlled and used judiciously for maximum effect, a process which requires elaborate equipment and a high degree of skill. Thus the processing conditions are as important as what goes into the oven.

Don Gubser's experience with the vagaries of processing came early on, just after the first reports of the new ceramics surfaced.

I had been assessing some Russian efforts, and this postdoc comes in to tell me that there was a report of a high-temperature superconductor that I ought to look at. I said, What's new? It happens all the time. Well, anyway, he said it really wasn't all that convincing, so I didn't pay much attention to it. But other people started saying they thought there was something to it, and they showed me the preprint from the Japanese. I looked at it and it had R equals zero, and a Meissner. I said, My God, this is real. So I dropped all of my administrative responsibilities, and I grabbed a ceramist to do the processing, and I said, We gotta make this fast, I think it's finally here.

Well, we started making it, and the ceramist told us we didn't know a thing about ceramics. I said, Well, the prescription is right here in this report. He said, But it's written by physicists. I said, I know, but can't we follow it anyway? And the ceramist says, They don't know how to make ceramic materials. So I said, OK, you make it for us your way. He did, and it didn't go superconducting. He tried it two or three times, and it didn't work. So my colleagues and I came in over the weekend, when the ceramist was gone, and we decided to follow the prescription, even though it was supposed to be wrong. Next day we had a superconductor.

Our processing is much better now, and we can get the properties right so they're reproducible. We know enough about the process and the conditions, and so we can always get to reasonable quality with 95 K materials. We have a lot more instrumentation now to do things that we had done by just pure happenstance. We've put eyes into the ceramists' methods. We put a resistor, say, into it so that as you go up and down the temperature scale, you monitor the differences as you see them, and that's called differential thermal analysis. Once the ceramists make the stuff, we go to an electron microscope and see what that little glitch means insofar as the microstructure is concerned. We go to the data acquisition system, the computer, and find out where the atoms are, what the crystal structure is, and we ask, Did the electrons go the wrong way?

This feedback, which twenty years ago could not have taken place, is almost standard to how we operate today. I know the reasons we were missing superconductivity back then was because we weren't taking care to monitor the positions of the atoms as we were processing. When we went in there that weekend, I spent more time on the x-ray machine than anyplace else in the lab. It was process, x-ray, process, x-ray, before we bothered to check for superconductivity.

But even with such material analysis, and even though a high school student can whip up a disk of passable superconducting ceramic, there's still a formidable difference between making an oxide superconductor and being able to pin down the theory behind this new curiosity of the physics world. Nobody has done the latter yet.

Not that there aren't plenty of ideas running around the labs. Sometimes they come almost as fast as the bursts of electrons in the gigantic research tools that the experimentalists and the theorists are using to get them some answers to the big question. Speculation and disagreement come with the territory. "Does Magnetism Hold the Key?" asks, the British journal *Nature*. One group insists that chains of copper and oxygen atoms are what make the new ceramics superconduct; another group says no, it's planes and layers, but absolutely not the chains. Then someone else finds out that another ceramic superconductor contains both the chains and planes of copper oxide. Somebody says it's the yttrium-barium combination that does the trick; somebody else says it's the copper oxide component. When I asked a theoretical physicist at Brookhaven whether his current idea held in light of a suggestion by a colleague that seemed to disprove it, he shrugged and cited Mark Twain's comment that difference of opinion was what made horse races.

"I read an article somewhere," Donna Fitzpatrick said, "that most of the theorists are keeping four or five theories going and that every other week they'll write a paper saying maybe it's this or maybe it's that, hoping all the while that someone will hit on the right one. When that happens, someone else will be able to say, 'Well, I thought of that didn't I?' I really don't think anyone else is committing to any one thing." Confirming that view was Du Pont solid-state chemist Arthur W. Sleight. "There's at least one theory for every theorist," he told a recent American Chemical Society news conference, "and some theorists have a package of theories in the hopes one of them may be right."

Some scientists are saying, perhaps unfairly, that the experimentalists have gone blindly galloping far ahead of the theorists, and that the theorists are caught up in a Cinderella game that requires them to fit their pet notions, their glass slippers, to the whole range of superconducting mixes that have cropped up. While it is true that the pace at which the new oxides have been accumulated seems to have left the theorists in the dust, this is not to say that all of the theorizing is blue sky. Scientific theories are, after all, intellectual tools essential to the discovery of new facts. Without them, the experimentalist—often thought of, unfortunately, as a mere laboratory tinkerer wholly isolated from the ominous-sounding "theoretician," the high-powered mathematician of the bunch—would be as misguided as a ship's crew running without compass or rudder.

One noted physics textbook for beginners addresses the issue squarely:

> Benjamin Franklin and Mme. Marie Curie were experimental physicists. Isaac Newton and Albert Einstein were theoretical physicists, perhaps the greatest. In the earlier days, the tools, both experimental and mathematical, were so simple that a single man or woman could become skilled in the use of both kinds. Isaac Newton not only made the thrilling experiment of breaking sunlight into colors with a prism, but actually invented for his own use one of the most useful forms of mathematics, the calculus. Franklin contributed to electrical theory. Nowadays some of the tools are so complex that few physicists are versatile enough to become masters of them all. But whether theorists or experimenters, the people who build physics are all physicists.

That may be another way of saying that the experimentalists who are manipulating all of the new oxides are not just groping about in a shed full of ceramic mixtures, nor are the theorists merely scribbling on blackboards and talking to themselves. Both groups have been working closely together—and they're not all physicists. Chemists, materials scientists, electrical engineers, and even medical doctors are deeply involved in superconductivity research these days. Team effort applies unquestionably as scientists explore the materials and mechanisms of the new superconductivity, just as it did when Bardeen, Cooper, and Schrieffer put their talents together to explain conventional superconductivity.

What, then, are the scientists finding out about how this new superconductivity works?

Before we try to answer that, a couple of things have to be kept in mind. First, as an expert in the field put it recently, ceramics do not live in the stone age, they live in the technological age. Their microstructure is intricate and unlike that of other, conventional superconductors. The other point to remember is that while the BCS theory applied elegantly to all the earlier known superconductors, it does not seem to apply entirely to the new ceramics and to their behavior at high temperatures—and indeed may not be applicable at all.

Let's examine the first point. The new oxide ceramics are no mere lumps of clay, nor of the same blend that goes into a dinner plate. They are heterogeneous materials, jumbles of tiny grains welded together but with impurities stuck here and there, causing boundaries. As the grains mesh, their ability to conduct electricity grows; if they do not mesh, current flow is lessened, and it is lowest at the area of the grain boundaries.

It is important, however, to get down to another level of magnification, that of the material's crystalline structure. Crystals are formed when a chemical element solidifies into three-dimensional geometrical shapes—triangles, parallelograms, hexagons, to name a few. These shapes—which may be likened to Tinkertoy arrangements, or to stacks of eggs in cartons, with each egg an atom—are the lattices we spoke of earlier in our discussion of ordinary electrical conductivity; they are what give each element its own special internal structure. All crystals of the same substance have identical angles between their faces. For example, common salt dissolved in water will always leave salt crystals in the form of cubes when the solution evaporates; a solution of blue vitriol, a substance with a number of industrial uses, including electroplating and textile dyeing, will form rhombic (loosely, diamond-shaped) crystals of copper sulfate on evaporating. A diamond is crystallized carbon, as is graphite, the "lead" in our pencils.

Crystalline materials, however, need not exist as crystals—all metals, for example, are crystalline, even though one doesn't usually see them as geometric structures. What is common to all crystals is that they are formed from the regular arrangement—the crystal structure, or lattice—of atoms, ions, or molecules. That regularity is the reason why an object, when broken, separates cleanly along the lines laid down in the lattice. So crucial is the alignment that when it is changed, so too is the material. When, for instance, steel is heated, the geometry of its carbon and iron atoms shifts. If the steel is cooled slowly, it reverts to its original, crystal structure; cool it too quickly, however, and a wholly different, more brittle atomic configuration results, as occurs in so-called high-carbon steel.

Something like that can go on in the new ceramic superconductors. In the yttrium-barium–copper oxide material, for example, if annealing—the heat treatment used to soften a material and make it more workable, and to relieve internal stresses and instabilities—goes on for too long, the ceramic begins to decompose; if the annealing step is too short, it doesn't superconduct.

It is here that oxygen, the most abundant of all the elements in these ceramics, plays its critical role. A component of a great number of organic and inorganic materials, it forms oxides when combined with all elements except the inert gases. The rate of that reaction, known as oxidation, varies from element to element. Oxygen can also intensify a reaction; when mixed with another gas in a welding torch, it produces a far hotter flame than can be obtained simply by burning gas in air. In liquid form, it can supply a mighty driving force to propel rockets and missiles. In the new ceramics, it fuels superconductivity. "We're learning a lot more about oxygen in these ceramics," Merwyn Brodsky observed. "It's all very, very fascinating. It turns out you want as much of it as possible in that structure. It creates the chains."

The chains Brodsky speaks of are one-dimensional, alternating linkups of copper and oxygen atoms that form the latticework of the ceramic crystal, a sequence that many believe is the key to high-temperature superconductivity. In the case of the yttrium-barium–copper oxide material, these chains run parallel to warped, two-dimensional planes composed of copper and oxygen atoms and separated by the copper-oxygen chains. Alternating ions of barium and yttrium hang inside each three-dimensional "compartment," or unit cell, formed by the planes and chains. It is through these layers that electricity is superconducted.

Even though the yttrium and barium (and any of the other elements that show up in the various mixes) are important—they donate electrons for Cooper pairing and may simply act as glue to hold the structure together—it is the copper-oxygen bond that seems to form the hot wire of this rig. Since the distances between the copper and oxygen atoms in the layers are not great, electron transfer, and thus the flow of electric current between them, occurs fairly easily.

Scientists looking into this latticework also discovered a couple of other important things about ceramic superconductivity. First, they found that as the new materials were prepared, their crystalline structures changed drastically— from a basic, nonsuperconducting configuration formed when the ceramics were first put into a hot oven, to a superconducting pattern formed as the material was cooled down. (Specifically, when the crystal structure of the material is orthorhombic—having different crystal dimensions in all three perpendicular directions—it superconducts at about 90° K. When the crystal's structure is tetragonal— different dimensions in two directions—it does not superconduct.)

The other bit of information researchers picked up was that when the crystal was in its nonsuperconducting shape, it had six oxygen atoms per molecule in its chains; in its superconducting form, it had seven. Moreover, the researchers found that if they annealed their ceramic—the step that followed heating it in the oven—in an inert atmosphere, it lost oxygen atoms; when they annealed it under oxygen, the supply was replenished. The arrangement of

HIGH-TEMPERATURE SUPERCONDUCTOR

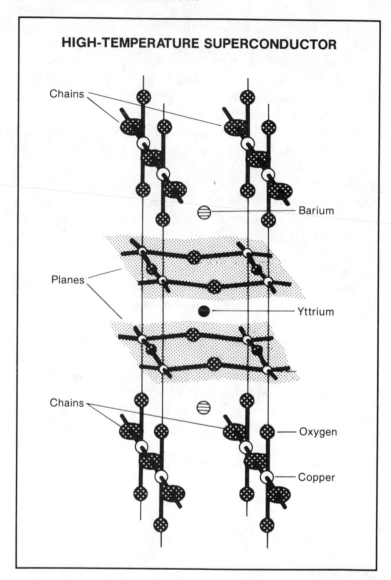

SUPERCHAINS AND PLANES. The atomic structure of ceramic super-conductors contains chains and planes of copper and oxygen atoms. Both chains and planes appear to contribute to high-temperature superconductivity in some of the new materials. (*Courtesy Argonne National Laboratory.*)

the oxygen atoms was what apparently forced the lattice to convert to a shape that would support high-temperature superconductivity: the more oxygen in the chains, the more rigid the order and the higher the superconducting temperature. Fewer oxygen atoms meant that the copper atoms were unconnected, like stepping stones spaced too far apart across a stream.

Getting to know this intricate structure was a high priority of scientists at several laboratories, among them Argonne, twenty-five miles southwest of Chicago, and Brookhaven, on the site of old Camp Upton, a training camp for U.S. soldiers during both world wars.

The two venerable laboratories' involvement was a given. Operated by the Department of Energy (or DOE, the umbrella over six national laboratories, and the agency responsible for, among other things, building and running high-energy particle accelerators), both had had a long and distinguished history in basic and applied research in the physical, environmental, and biomedical sciences, and in energy technologies. Between them, they also had the super sophisticated instrumentation necessary to explore what, a few years ago, would have been unexplorable. These powerful, expensive devices possess names as imposing as their bulk: the Alternating Gradient Synchrotron, the High Flux Beam Reactor, and the Tandem-Linear Accelerator System (ATLAS, for short). And they had the manpower and the money: Brookhaven with 3,200 employees, 1,800 collaborators and other visiting researchers, and an annual budget of about $258 million; Argonne, with 4,100 employees and an annual budget of about $200 million.

At both labs, interest in superconductivity predated the high-temperature superconductors. In 1962, Brookhaven was investigating the use of superconducting magnets for accelerators, concentrating on niobium-tin, despite its brittle nature. A high-field solenoid made from niobium-tin was built in 1964 and was one of the first magnets to result from the lab's efforts. In 1968, the lab had a seven-foot bubble-chamber magnet (a device for detecting ionizing radiation), the largest air-core superconducting magnet in operation. A few years later Brookhaven had a pair of superconducting magnets in one of its big machines to bend a proton beam on its way to the chamber, and that, at the time, was the largest superconducting accelerator magnet in operation, and the only one used for primary beam transport. Today, the origin of most superconducting accelerator magnets can be traced to the so-called CBA Palmer magnet, named for the man who designed it, Robert Palmer, former associate director for high-energy physics at Brookhaven. The lab can also point proudly to another association: Long before Alex Müller discovered superconductivity in metallic oxides, he had studied the perovskite mineral, whose three-dimensional structures resembled the new high-T_c materials. In fact, in 1969, Müller collaborated with Brookhaven scientists, working with the High Flux Beam Reactor at the lab to investigate the makeup of a perovskite compound.

For its part, Argonne had done more superconductivity-related research than any nonindustrial laboratory in the nation. Indeed, as long ago as the 1960s, the laboratory designed and built the first high-energy-physics particle detector to use a large, superconducting magnet. In the 1970s, it constructed a 40-ton magnet for a joint research project with the Soviet Union, and a second, weighing 180 tons, ranks among the largest superconducting magnets ever built. Argonne also developed, at around the same time, a high-temperature superconducting quantum interference device (called a SQUID) to measure magnetic fields as small as one ten-billionth of the earth's magnetic field; and a superconducting tunnel-junction transistor, an ultra-sensitive device that amplifies electrical signals up to four times and has potential uses in superfast computers and energy technologies that use superconducting magnets. In 1987, President Reagan named Argonne a national center for the study of applications for the new materials; more specifically, the DOE selected it, along with Brookhaven and the Ames Laboratory in Iowa, to develop practical conductors from the ceramics in the form of wire and thin films within five years. Argonne was the first American research organization to make a wire out of the new ceramic materials, and the first to put a current through it.

It soon became apparent, once the structure of the yttrium compound was bared, that either or both of two central features might account for superconductivity at those record-high temperatures. One was the puckered, two-dimensional plane of copper and oxygen atoms
ilar to the flat plane seen earlier in the structure of another superconductor, made of lanthanum, strontium, and copper oxide, that became superconducting at around 40° K. The other was unexpected: the one-dimensional chain of copper and oxygen atoms, a sequence unknown in earlier superconductors. The challenge was fairly clear to both theorists and experimentalists. Were the planes or the chains responsible for superconductivity above 90° K?

Over the next few months, several experiments at Argonne seemed to confirm that although superconducting currents flow through both chains and planes, the chains were the key structural feature that produced high-temperature superconductivity in the yttrium-barium–copper oxide.

It all sounds logical enough, but as with the wonderful one-hoss shay that was built in such a logical way, scientific theories don't necessarily always last and, in fact, often fall apart all at once. At this writing, the new bismuth materials were posing a puzzle for the researchers. They have the layered sheets of copper and oxygen atoms, as in the other high-T_c superconductors, but they seem to lack the chains of atoms. The lanthanum superconductor caused some more head scratching: its structure was sometimes less symmetrical than orthorhombic. It could well be that these materials are an entirely different class of high-T_c materials based on copper-oxygen planes. It may also mean that other undiscovered materials and mechanisms are waiting somewhere in the periodic table of elements.

But even though crystallographers have pulled the ceramic structure apart, can track the path of a current along their chains and planes, and can speculate as to *where* superconductivity actually takes place, this doesn't yet tell us *why* and *how* it all happens. Still lacking is a theory to explain the mechanism of the new superconductivity, one as rigid as the BCS theory that explains so well low-temperature superconductivity in metals.

What helps make the formulation of a new theory so difficult is that the ceramic superconductors have such unusual physical properties. Their "dirty" nature sometimes gets in the way and complicates the measurement of their microscopic properties. They don't shine, even when polished, and reflect unusually little light; since light reflection is a clue to what electrons are doing inside a solid— particles of light called photons settle inside, react with electrons, and bounce back out to give the shine—the lack of reflected light means the ceramic electrons are behaving in some unconventional way. The ceramics can be turned into insulators fairly easily, simply by changing their oxygen content; some of them will even dissolve if left overnight in a glass of water. Because of their layered structures, they act like metal conductors only in directions parallel to the layers, a property known as anisotropy. In that sense, they are like wood, with strength along the grain different from that perpendicular to it.

Such peculiar behavior, rather than serving to deter theorists, has only challenged them more, and most feel that although a solid microscopic explanation of the new superconductivity may be elusive, it is not unattainable and could be in hand within five years. What form it will take, however, is literally anybody's guess at the moment. But there is agreement on one point: the pairing of electrons, as set forth in the BCS theory for low-temperature metallic superconductivity, still seems valid for the new materials, which means that it may not be necessary to search for larger groups of electrons to explain the phenomenon.

That pairs are at work in the new ceramics has been demonstrated by measurements of what is known as the Josephson effect, named after the physicist Brian D. Josephson, who observed it in 1961 as a graduate student at Cambridge University in England. The Josephson effect, like the Meissner effect, is one of the many properties of superconductors and is not only relevant to the mechanism of superconductivity but permits many of the electronic applications that researchers envision for the new materials.

What Josephson found, based on the BCS theory, was that if two superconducting metals are separated by a thin insulating barrier, like an oxide layer, it is possible for electron pairs to pass without resistance through the barrier (known as a tunnel junction). Several teams of scientists have already observed the Josephson effect in the new materials, among them Dr. Alan Clark's group at the National Bureau of Standards' Electromagnetic Technology Division in Boulder, Colorado, which detected it in the yttrium compound at liquid nitrogen temperature. To find evidence of the effect, scientists at the NBS used a technique devised by Dr. John Moreland called "break junction," which involves bending a bulk sample of the superconductor with an electromagnet until the sample breaks. This creates a tunnel junction between the split ends. Electrons flow across the junction, creating a tunneling current that reveals characteristic features of the superconductor.

According to measurements made with the technique, the size of the energy gap—defined in very general terms as the strength of the coupling between the paired electrons—in the ceramic oxides corresponded to the values predicted by the BCS theory. The NBS team also found evidence of a strong coupling mechanism that was much stronger than in metal superconductors. This was significant, since it seemed to indicate that some mechanism other than the phonons of the BCS theory was the mediator.

But just what causes the pairing in the high-T_c superconductors is unclear, and scientists disagree on the mechanism. Some researchers believe that the phonon-electron interaction is important, as it is in the BCS theory. But a growing number now believe that something else is at work. The intermediary, many feel, is some sort of magnetic, electronic excitation in the crystal lattice structure, a force created by such subatomic particles as excitons or plasmons.

Whether that turns out to be the case is still anybody's guess. At this writing, the pairing mechanism, the electronic glue, is still a mystery. It might well be that the ultimate explanation will involve a combination of mechanisms. Or it could be that something totally unheard of, something that has nothing to do with pairs of electrons, is going on in the new superconducting materials. Indeed, some scientists have suggested that the new ceramics are really new kinds of metals that carry electrical charges, not via electrons, but through other charged particles. As Princeton physicist and Nobel Laureate Philip W. Anderson put it recently, in the title of a paper he gave reviewing the history of superconductivity theory: "It isn't over till the fat lady sings."

8

ON A MORE PRACTICAL NOTE

How the high-temperature superconductors work is, of course, a vital scientific question, and for the researchers who are losing sleep over it, an answer would be an end in itself. Let others pursue the search for superconductivity's practical applications. As for themselves, they are committed to the dictum of Max Planck, whose quantum theory ushered physics into the modern era. "Scientific discovery and scientific knowledge," said Planck, "have been achieved only by those who have gone in pursuit of it without any practical purpose whatsoever in view."

That is one view. Another showed up recently in a Japanese comic book on superconductivity. The book, published by Goma Shobo, is one more in a line of popular Japanese comics whose goal is to present abstract material in a light vein. It is all about an industrial spy, a former physicist who lost his research funding when his work failed to produce results. The spy explains superconductivity to a college friend turned businessman, and the businessman, quite naturally, sees dollar signs and hires his friend to infiltrate laboratories working on new high-T_c

materials. Technical details about superconductivity
abound, mostly presented in the form of a university lec-
ture overheard by the spy. The reader learns what superconductivity is, all right, and about the Meissner effect and
the importance of yttrium-barium–copper oxide. But there
is much more, as a review of the book in *Scientific American* saw it:

> The comic book also does what most scientists
> studiously avoid: it discusses in grand terms the
> presumed payoffs of the new superconductors.
> The story does not shy from the realpolitik of
> science either. At one point, the spy explains to
> his friend that unless Japan develops high
> temperature superconductors, the U.S. could put
> Japanese researchers out of the race by banning
> exports of liquid helium.
>
> Later, the spy proclaims, "This is not the time
> to conceal superconducting materials or lock
> them up in secret rooms."
>
> "You are too naive," his friend retorts. He adds
> that American politicians were already advocat-
> ing secrecy, aimed at locking out the Japanese.
> Eventually, both the spy and the profit-hungry
> businessman get their just deserts: the business-
> man finds he cannot patent a trivial supercon-
> ducting application and the spy falls into the
> arms of a pretty girl.

A bit farfetched, perhaps, but it would be certainly naive
to deny that the IBMs and the AT&Ts, the GEs and the Du
Ponts, the Westinghouses and the rest of them are not in it
to commercialize the new technology even as they break
new ground on the basic research side. And for the most
part, they admit it.

Praveen Chaudhari of IBM addressed this matter before
Congress:

> Are we concerned that one of our competitors
> could be the first to exploit this new technology?
> Absolutely. But let me assure you that we are

working diligently and resourcefully to determine if it is possible to manufacture a computer
using superconducting materials. We feel we've
already made significant progress toward that
end, and we believe that we have the necessary
resources and dedication to continue our efforts.
To IBM, the recent breakthroughs in superconductivity research represent an exciting new
technological opportunity, and we will vigorously continue to determine its feasibilities and
exploit them to the fullest.

Chaudhari, quoted recently in the *New York Times*, was
also not averse to engaging in a bit of reverie about superconductivity:

You don't even have to make cars. You could
make little gizmos, you could put on a pair of
special shoes and make little tracks along which
you as a human being could push yourself and
keep going. Nothing to stop you, right?

I can see the whole transportation system being very different. At airports, instead of these
long conveyor belts we have, you could get onto
one of these platforms that are levitating and just
stay on it while it takes you around.

Chaudhari also envisioned floating furniture guided by
wires embedded in the floor: to get a table or a chair to rise
and sink, designers would use small electromagnets that
could be controlled with a simple dial. "It just pops up,"
said Chaudhari, "and the strength will determine how high
or how low it will go."

Over at Bellcore, John Rowell was a bit more restrained,
but his interest in the practicality of it all was evident. "We
do feel," he said recently, "that a very unusual scientific
event has occurred which may have some technical implications. It's a whole new ball game, parallel to someone
suddenly discovering some new kind of magnetism. And if
there is something drastically different that's going to arise
as a technology, we feel that Bellcore has got to be there.

It's an insurance policy, or maybe it's more like a spy on a mission who has to gather vital information he can use later on." (No, Rowell hadn't seen the Japanese comic book.)

Added Merwyn Brodsky, "Maybe we're driving the field too hard. But it's hitting at a time when the U.S. industry is asking what we can do to stay high-tech. This is it."

As research institutions plunge into the superconductivity sweepstakes, their pride, along with their willingness to talk about commercial applications, is strong. Few of the practical-minded, in fact, seem to be hiding their lights under bushel baskets, and although occasions arise when companies are reluctant to divulge specifics about some superconducting device or manufacturing procedure ("Naw, you don't want to visit our lab," one scientist told me. "You see one, you've seen 'em all"), it is fairly easy for writers to secure interviews with most of the people working in the applications field.

At Brookhaven, that pride shows in a house publication that proclaims that the lab "has pushed the technology of superconducting magnets beyond any other laboratory in the world," and that it is now working on the design of magnets for the Superconducting Supercollider, the 20-trillion–electron-volt proton-beam accelerator that will, it is hoped, provide an unmatched look inside the world of elementary particle physics and beyond, into events that took place in the cataclysmic, chaotic split seconds when the universe was being conceived.

Brookhaven's pride is justified, for if built, the SSC will be the single most awesome example of superconductivity's powers ever devised.

And Don Gubser of the Naval Research Laboratory boasts shamelessly of the navy's strong historical and continued interest in superconductivity. "If you want to talk about an agency that has full breadth of potential applications, intent, and support for superconductivity, you won't top the navy. NASA gives us money just to teach them to make the materials."

Where will it all lead?

JUNCTIONS, SQUIDS, AND OTHER ELECTRONIC DEVICES

At the National Bureau of Standards laboratory in Boulder, Colorado, researchers regularly pull all-nighters at their beam lines on the Synchrotron Ultraviolet Radiation Facility–II, trying to pin down the electronic structure of batches of superconducting ceramics. "I can't see keeping away from this," says one of the scientists, Dr. Richard Kurtz. "This is so hot that I can't imagine anything else would be as exciting."

In the same lab, physicist James Zimmerman, internationally known for his work on superconducting electronic devices and the refrigerators used to cool them, carefully adjusts one of his latest gadgets, a superconducting quantum interference device (or SQUID) that is a high-temperature version of the most sensitive existing device for measuring magnetic fields. Zimmerman has been lured out of retirement by the superconducting heat wave to perform this experiment, and it pays off: bathed in swirling white vapor, the SQUID, made with a sample of yttrium-barium-copper oxide, will operate at temperatures up to $81°$ K, $4°$ K above the temperature of liquid nitrogen, and may well be the first practical superconducting device operating in the relatively cheap coolant.

National Bureau of Standards scientists were studying superconductivity long before it became fashionable. Committed to turning a laboratory curiosity into useful technology, NBS researchers had, since the 1960s, achieved other record-setting performances in superconducting electronics by building devices that exploited the unique measurement capabilities of superconducting materials. Among these were microwave "mixers" that detected single photons, particles that travel at the speed of light; superfast analog-to-digital converters; and arrays of Josephson junctions, a series of 14,184 highly accurate electronic devices that can convert a precise frequency to a precise voltage.

Buoyed by such developments, along with the knowledge

that the conventional superconductor industry registers annual sales of around $200 million, it is little wonder that scientists studying and measuring the properties of the new superconductors foresee great commercial promise for them in electronics.

With the role it's played in the development of computers, communications systems, and medical and scientific instrumentation in particular, electronics has already accomplished what an old Italian proverb says about wine: it has worked more wonders than a churchful of saints.

The modern era of electronics began in 1906 with the invention by American radio engineer Lee De Forest of the vacuum tube, the squat glass and metal bulb used in early radios. His creation was really an adaptation of a bulb invented earlier by English electrical engineer John Ambrose Fleming and which served as a rectifier to change alternating current to direct. Fleming's tube—he called it a valve because it acted as such in controlling the flow of current—took advantage of the ability of electrons to flow out of metal filaments heated in a vacuum.

Fleming put but two electrodes in his tube (making it a diode), a filament to emit electrons, and a cylindrical plate wrapped around it to collect them. When the plate was positively charged, it attracted the filament's electrons, establishing an electrical circuit loaded with current.

De Forest improved the device by making a simple change. He inserted a third electrode of fine wire (called a grid) between the plate and the filament; the grid attracted electrons and sent them speeding more quickly to the plate. Even small changes in the voltage applied to the grid in this new tube, now known as a triode, drew out large numbers of electrons and hence more current.

This capacity of the grid to control and increase current flow now gave the vacuum tube the power to amplify electrical signals. It was this variation that made obsolete the headsets used by early radio operators and opened the way to a whole new world of efficient communication by radio, telephone, telegraph, television, and public address and motion-picture sound systems. For more than forty years, the vacuum tube, because of its versatility and economy, controlled the majority of electronic applications—radar, computers, and weapons systems among them.

Then came the bulky tube's first serious competitor, the transistor, a mighty mite that was the electronic equivalent of the mechanical valve and could control large quantities of electrical current with startling efficiency. About the size of a pencil eraser, mechanically rugged and versatile, and with a longer life span than tubes, this oscillator (a device that produces alternating current), rectifier, switcher of signals from one circuit to another, and amplifier of electrical power could do everything the vacuum tube did and more. It could also do it more cheaply, more dependably, producing relatively little heat, and using far less electricity. Without the transistor, there would be no high-speed computers, no pocket calculators, no communications satellites.

Transistors are made of tiny slices of the crystals of semiconducting materials, such as silicon or germanium, to which wires are attached. In the early versions, two of the three electrical contacts—comparable to those in the vacuum tube, but miniaturized—were fine wires pressed into the surface of a semiconductor crystal in close pairs to create areas where current would flow only in one desired direction.

Because of their fragility and occasional unreliability, the point-contact electrodes were eventually replaced with three layers of adjacent semiconducting surfaces, each of which corresponded to an element in the triode vacuum tube: the emitter layer (for the heated filament which is the source of electrons), the base (for the grid that controls the electron flow), and the collector, for the triode plate that receives the electrons. The areas where the layers join one another are called junctions, and transistors made in this way are known as junction transistors.

Eventually, manufacturers fixed several transistors onto single pieces of semiconductor material, a technique which led to the so-called integrated circuit. These miniature electronic circuits, the familiar chips, may be as small as 1 millimeter square, as in the logic circuits that do the arithmetic and control the operations in a computer. A single integrated circuit in a programmable pocket calculator typically carries some thirty thousand transistors on a minuscule silicon chip. Large-scale circuits, perhaps 8 millimeters square, can contain a million or more transistors and other electronic elements.

Because the function of semiconductors, like the new superconducting ceramics, is dependent on the way their crystalline structure is arranged, transistor makers can control the function of the devices by deliberately adding small amounts of certain impurities to the crystals. Such impurities control the flow of electricity through the device either by providing extra electrons or by not supplying enough.

But transistors, as tiny as they are, were not the last word in miniaturization. For all their amplifying ability, they are still dependent on ordinary conductivity. The discovery of superconductivity opened the door to important developments in circuitry, and a number of new devices emerged.

One of these was the cryotron, a fairly simple switch made of two superconducting metallic wires, one coiled about the other. Made in the form of thin films and locked into the circuitry of a computer, cryotrons—because they are superconductors and require so little electricity, only start-up power—made possible much smaller computers. Moreover, their incredible signal-switching speed—measured in billionths of a second—has been a boon to computer designers striving to make faster machines.

The Josephson junction is one such ultrafast superconducting switching device. Josephson junctions, which until recently operated only at liquid-helium temperature, are traditionally made of niobium-tin or niobium-germanium and are really simple connections between superconductors. They can do everything vacuum tubes and transistors do, but a lot faster.

Josephsons come in a variety of sizes (all a few micrometers and smaller) and shapes. Some are blemishes in thin superconducting films (look hard and you still won't see the microbridge, a short, narrow constriction that controls the device's applications) and are thinner than the film itself. Others are snippets of fine-pointed superconducting wire in contact with blunt, superconducting posts, or they are drops of lead-tin solder hardened around a niobium wire. One model is called a superconductor-semiconductor-superconductor, another the superconductor oxide–normal metal–superconductor.

Because they are connectors between superconductors, the junctions and instruments which incorporate them are traditionally used to study the Josephson effect (the ability of a current to flow with zero resistance through an insulating barrier separating two superconducting materials), to define highly accurate voltage standards, and to measure magnetic fields. They are also, like the vacuum tube and the transistor, capable of amplifying electromagnetic signals and of switching those signals from one circuit to another. Moreover, Josephson junctions are able to perform their switching functions, which include switching voltages, with incredible speed—around ten times faster than ordinary semiconducting circuits—and they do so without using up a lot of power, since they are superconducting devices. These are distinct advantages in a computer, which relies on brief, on-off electrical pulses to carry information in digital form. The shorter the pulses, the more data that can be carried. Ordinary, nonsuperconducting wire has a tendency to waste the electrical pulses, something that a superconductor, of course, does not do.

Since computer speed is dependent in part on the time required to transmit signal pulses within and between different circuit elements, switches can be a limiting factor. The junction devices' phenomenal switching speed—along with their low power consumption, which allows them to be packed tightly together without generating much heat—makes them ideal candidates for use in the logic components of a superfast, and much smaller, computer. Current supercomputers—like the Cray 2, manufactured by Cray

Research, Inc.—are the fastest processors in the world, generally sending pulses in a billionth of a second—a nanosecond. They are used to design aircraft, drugs, or nuclear weapons, to crack codes, track missiles, even forecast the weather.

But for all their awesome number crunching, they are big, and the electrical resistance in their circuits generates enough heat to require expensive cooling systems. Even when a relatively light current flows through the computers, so many microscopic transistors and other elements are packed onto the chips that heat accumulation from electrical resistance threatens to melt them. Another barrier to running an ultrafast computer is the fact that electricity cannot move faster than the speed of light. Even the briefest delay when a computer idles, waiting for a signal to move a hair's breadth, slows the computer's speed, and this can mean that millions of operations will go undone. Closer packing could solve that, but the heat such a strategy creates stands in the way.

If and when a Josephson junction computer is built, the junction's size and low power dissipation would allow manufacturers to put more guts and gas into their machines. Their cycle times—the time required for a chip to perform one task—would be substantially shortened. Such a computer might, in fact, fill a cube only 2 inches on a side and operate more than fifty times faster than the best that are available today. No mean feat, considering that the world's first all-electronic computer, ENIAC (for Electronic Numerical Integrator and Calculator), covered some 1,500 square feet of floor space at the University of Pennsylvania, where it had its maiden run in 1946, was jam-packed with some twenty thousand vacuum tubes, and weighed in at more than 30 tons. Moreover, its computations were measured in seconds—not a nanosecond, a picosecond (a trillionth of a second), or a femtosecond (a quadrillionth of a second), the measurements computer designers are accustomed to shooting for today.

Scientists in the United States and Japan have already developed all the circuit elements required for a Josephson junction computer, but its design generally would rely on metallic superconducting materials. Even if they manage to get these conventional superconductors to work successfully on a chip, however, there will be difficulties. Chief among these is the interconnect problem, a limitation on the speed of an integrated circuit by the electrical connections that connect the transistors on a chip, and the chips to other chips. These contacts have a fairly high electrical resistance, and though such resistance causes heating in any device, it is especially troublesome in superconductors since even a little heat can raise their temperature enough to stifle or destroy superconductivity. The interconnect dilemma is serious enough in computers, because heat generation limits the number of circuit elements and consequently the ultimate speed of the machine. It is even more of a concern in magnet applications, where large contact areas are required to handle very high currents.

The new superconducting ceramics offer a way out, if they can be made in some workable form, probably a thin film, and so long as the circuits can be operated at around 80° K or higher. The oxides could then be used on the circuit chips themselves to connect transistors and other devices and to make connections between chips. There is also the possibility that the superconducting properties in the new materials can be manipulated by, say, moving oxygen in and out as needed, or that the ceramics will turn out for some reason to be more suitable than conventional superconducting materials for use in transistors.

But getting the new superconductors to stick to a chip presents some unique problems and there is also the difficulty of applying the superconducting material to a useful substrate.

Chipmakers would like to be able to use a silicon dioxide, the material from which virtually all the world's microchips are made. But until very recently, it had not performed well with the new ceramics because the superconductor's components mix with those of the silicon, causing a rearrangement of elements. The resulting material may no longer have the yttrium, barium, and copper at the approximate 1:2:3 ratio required, which would explain why superconductivity is destroyed. Scientists at the General Electric Research and Development Center in Schenectady were the first to report the successful operation at liquid nitrogen temperature of an yttrium-barium superconducting film applied on silicon.

The key was the insertion of a zirconia buffer layer between the silicon substrate and the superconducting film on top of it. (Zirconia is a white crystalline compound used as an insulator in enamels and as an electrolyte in fuel cells.) The buffer, deposited with superconducting film onto the silicon by electron-beam evaporation, served as an effective barrier, preventing the elements from intermingling during the annealing process.

Zirconium oxide has also been used as a substrate by itself. Researchers at Cornell University evaporated some of the yttrium superconductor with beams of high-energy electrons, deposited the vapors onto bits of the zirconia, and then etched a circuit pattern a fraction of an inch long. Not only did the superconductor film carry current of around 1,000,000 amps per square centimeter, but it conducted electrical impulses as brief as ten to fifteen-trillionths of a second without distortion—impossible with conventional materials—and at very high levels of current. Those incredibly short pulses raise the distinct possibility that an enormous amount of electronic data, not only in a computer but in a telephone line as well, can be transmitted via the new superconductors at ultrafast speeds.

The modern era of telecommunications has begun to depend on bundles of optical fibers—thin, flexible tubes of glass and plastic—to send information over long-distance telephone networks. Also used in medical instruments to examine various body cavities, such as the bladder and intestines, a fiber-optics system converts electricity to pulses of light that carry the data—much more of it, in fact, than ordinary wires, which have to use pulses of electricity.

Just how much data a fiber-optic system is capable of handling was experimentally demonstrated recently by

Bellcore scientists, who set a record for what could be transmitted over a single, hair-thin fiber: the equivalent of 27 billion bits of information (called gigabits). By using eighteen lasers to broadcast information throughout a star-shaped network at different wavelengths, the researchers were able to make their system perform complex interconnection functions, and endow it with the transmission capacity to convey 400,000 simultaneous conversations, or enough capacity to transmit in 1 second the contents of 2,500 standard-size textbooks of 400 pages each.

If the superconducting film developed at Cornell is ever put into a telecommunications system, those stats would pale in comparison. Fast as they are, fiber-optical pulses must work in three steps: conversion of electricity to light and back to electricity, a process that is relatively slow. A superconductor transmitting electrical pulses alone could, theoretically, carry more than a trillion bits of information per second—enough to handle 15 million simultaneous phone conversations, or to transmit everything in print in the Library of Congress in a couple of minutes!

But let's return to Josephson junctions for a moment. Even though a computer made of these incredible instruments has not yet been built, the junctions themselves, as we have said, are in use, fabricated of conventional superconducting materials. They are also beginning to appear in devices run by the new superconducting ceramics.

One instrument in which the standard Josephson junction plays an important role is the SQUID (superconducting quantum interference device). Incredibly sensitive magnetic field detectors, these devices can measure forces as small as one ten-billionth of the earth's magnetic field and are widely used to monitor brain waves and voltage, in geological prospecting, in military surveillance such as undersea submarine detection, and in fundamental physics studies. Like the Josephson junctions, SQUIDs are not much to look at, many of them being made from niobium components and plain old thin films trailing current and voltage leads.

But it is not their simplicity that concerns researchers

trying to improve them. It is that they can only operate at 4° K, the temperature of liquid helium. That may soon change. Several groups have managed to construct SQUIDs from the new high-temperature superconductors, which means they can work in the liquid nitrogen range. IBM was probably the first to pull it off. In April 1987 the company announced that it had made a SQUID from two thin-film Josephson junction devices, the whole affair about one-hundredth the thickness of a human hair. It went superconducting at up to 68° K. (Although liquid nitrogen boils at 77° K, it can be effectively used at 68° K by reducing its pressure.) It was in May of that year that scientists at the National Bureau of Standards made a similar device and got it to operate at 81° K—4° K above the temperature of liquid nitrogen. Not a heat wave, by any stretch, but one that would, it is hoped, have far-reaching implications for a whole new generation of electronic devices.

9

SHIPPING OUT WITH SUPERCONDUCTIVITY

IT WASN'T REALLY a ship, this crudely shaped hull resting on blocks in a dusty laboratory storeroom at the University of Mercantile Marine in Kobe, Japan. Just a 12-foot, wooden-hulled model that looked as if it had been knocked together out of spare parts by a steam-fitter working from blueprints for an antiquated carnival ride.

It had no engine, no propellers, and no rudder, and its top speed, during its brief shakedown cruise in a tank full of salt water, was only 1.5 knots, which left a wake that even a duck wouldn't be proud of. "But," shrugged its developer, physicist Yoshiro Saji, "it moves, and that's a beginning."

The *ST-500*, as Saji's creation is called, is obviously no ordinary vessel. With wires and plumbing and valves jutting from its blue and white hull, it could well give the art of shipbuilding a bad name. But though it lacks flash and dash, the *ST-500* could have a dramatic impact on the speed and fuel efficiency of future ships as scientists learn more about applying superconductivity to ocean transportation. Although magnetic levitation rail systems have captured most of the media's attention in recent years,

superconducting ship-propulsion systems have been studied since the 1960s, and several prototypes were built and
tested.

There are two concepts: Saji's, and one the U.S. Navy
has been interested in since 1969 or so. Both rely on superconductivity, but they work in totally different ways.

What gives the Japanese vessel such potential is its
unique power source—not oil or coal or sail, but the same
driving force that makes the rotors in electric motors spin,
electromagnetism. The working principle of such power is
something called Fleming's left-hand rule, named for John
Ambrose Fleming, who created the precursor to Lee De
Forest's vacuum tube. The rule says, in general, that a
magnetic field plus electric current produces a linear force.
As every high school physics student should know—and
as we noted in our discussion of magnetism—when an
electric current travels through a wire or any other conductor suspended in a magnetic field, it produces an electromagnetic force that pushes against the field of the magnet.
In an electric motor, current flowing through the armature
reacts in the same way against a magnetic field generated
by electromagnets, creating the torque, or push, that turns
the rotor.

The same thing happens in Saji's magnetic ship. It is, in
effect, an ingenious version of what is essentially a DC
motor. Except that there is no motor and no rotor. Indeed, it
has no moving parts at all. Instead of wire, the conductor

FLEMING'S LEFT-HAND RULE

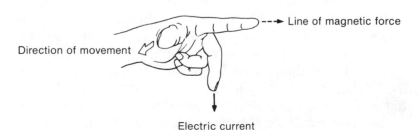

A demonstration of Fleming's left-hand rule.

through which the electric current flows is the ocean itself. Seawater, unlike fresh water, contains a rich concentration of dissolved salts that are good conductors of electricity.

In a conventional ship, powerful internal-combustion engines fed by diesel fuel generate the power that turns screw propellers, which simply push the vessel through the water. In the magship, as the craft has been informally christened, superconducting magnets arranged along the hull beam a powerful magnetic field into the surrounding water. At the same time, a generator sends an electric current into the water between electrodes attached to the underside of the hull. Because the current flows at right

ANATOMY OF A MAGSHIP

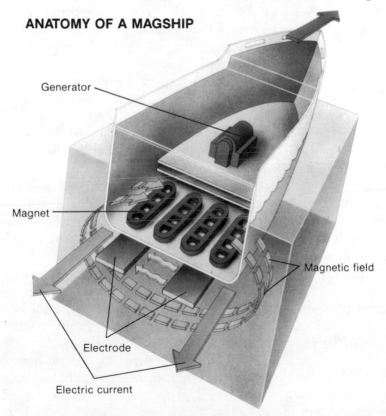

The Japanese Magship, shown here in a cutaway drawing, is propelled by magnetic thrust that relies on the principle of magnetic repulsion. In this instance, between the field from an onboard magnet and an electric current in the sea.

angles to the magnetic field generated by the superconducting magnets, electromagnetic force is exerted against the
conductive seawater, driving it backward and thrusting the
vessel forward. Simply by reversing the positive and the
negative charges of the electrodes, the ship can be made to
move in reverse.

The same principle underlies the maglev system that
will be used in the high-speed trains of tomorrow. The
train becomes a sort of flying carpet, levitated a fraction of
an inch or more over a tracklike guideway by electromag-
netics. But whereas the maglev train requires metal guideways and ground coils, the magship relies on conductive
seawater as a sort of fluid track. "The whole thing is the
simple principle of magnetic repulsion," Saji explained to
me. "But instead of using one magnet to repel another, as a
child might do when he plays, there is only one set of
magnets on board the ship, setting up the magnetic field in
the water. The repelling force is the water itself."

Saji had been grappling with the physics and engineering of magnetic propulsion for more than fifteen years, and
his blackboards reflect the arcane nature of the field:
they're covered with ragged-looking formulas used to cal-
culate such essential elements as the amount of electric
current needed in seawater, the density of the magnetic
flux amidships, and how best to use the superconducting
magnets to generate the greatest possible thrust. Those
superconducting magnets are the driving force that may
ultimately make possible the construction of full-sized
magnetic ships, submarines, research vessels, and mag-
netic-powered sea bases—floating structures that could be
used as virtually anything from oil rigs to oceanographic
research labs, or even midocean resorts.

But the idea that a ship could be driven by an on-board
magnet and an electric charge in the ocean was not Saji's. It
had been around for years, primarily the brainchild of
Stewart Way, an American engineer who worked forty
years in near obscurity for Westinghouse. Way had han-
kered as early as 1958 to develop an electric submarine
without propellers or jets; now in his seventies and semire-

tired in Whitehall, Montana, he is both pleased and astonished that his theories are finally bearing fruit.

Way chose submarines because they are subject to less drag than other vessels. Since they are not slowed down by surface tension and waves, submarines can take full advantage of the higher speeds promised by electromagnetic thrust (EMT). But in those early days, there was a seemingly insurmountable problem. A full-scale version of the kind of sub Way envisioned would have required magnets weighing more than 500,000 tons—about eighty times the weight of a Polaris submarine.

In the early 1960s, however, Way's vision became more practicable with the development of superconducting magnets. Enveloped in cooling liquid helium, they could circulate indefinitely and without energy loss the massive currents needed to produce intense magnetic fields. Convinced that the new magnets would one day make possible a full-sized electromagnetic submarine, Way decided it was worth building a smaller version with conventional magnets to test the basic propulsion principle. He left Westinghouse to teach at the University of California at Santa Barbara and promptly got his senior engineering class to help design an experimental magnetic sub.

With about sixteen hundred dollars' worth of storage batteries and an assortment of electromagnetic coils, aluminum tubing, and fiberglass-reinforced plastic, the group went eagerly to work and fashioned what they called the *EMS-1*. The torpedo-shaped craft was 10 feet long, 18 inches in diameter, and weighed 900 pounds. Launched in July 1986 in the Santa Barbara yacht basin, it ran silent if not deep, attaining close to 2 knots, 3 feet under the surface, for about 12 minutes. Way's excitement at the sea trial was toned down for the scientific report he presented later, but there was no mistaking his feeling that the sub's promise had been amply demonstrated. "In large sizes, this type of submarine would give exemplary efficiency," he wrote in the journal *Mechanical Engineering*. "A superconducting magnet would be used, and such a craft might be considered as an effective cargo carrier."

At the time, however, superconducting magnets were prohibitively expensive to operate, making electromagnetic thrust far less efficient than conventional power. Also, some scientists expressed concern about a potentially dangerous by-product: the chlorine formed when an electrical charge passes through salt water. They feared it could become a serious source of pollution if fleets of magships were operating on the high seas. (A new electrode material that gives off oxygen rather than chlorine has now been produced experimentally and may resolve the problem if the magship concept ever gets off the drawing boards.)

Enter a new class of lightweight superconductors, still expensive but much more efficient than earlier versions. Put them into the hands of the likes of Saji and his colleagues, who epitomize the celebrated Japanese aptitude for ingenious adaptation, and wait for the inevitable results. The first Kobe magship model, the *SEMD-1*, appeared in 1976. It took two years of experimenting, and then three months to build. "The new magnets were the key," said Saji. "We became the beneficiaries of Dr. Way's work when we built the world's first superconducting model." The *ST-500* followed in 1979.

Saji and his group are convinced that a full-sized ship should be next—and soon. His goal is a 100-ton magship in the next three years, and its development, he argues, would represent an important advance for the shipping industry, with some fairly obvious benefits. Because the magship would have no moving parts, it would be easier to build and maintain. It would move without noise or vibration, of particular importance for submarines and research submersibles, which often depend on silence to carry out their work of hunting and tracking enemy ships, and investigating undersea life. A conventional ship's propeller assembly produces drag as well as thrust, and in rough weather the props often race as the stern heaves in and out of the waves. A propless magship, on the other hand, relying on the sea for its power, could generate maximum thrust in nearly any weather. An icebreaker would find electromagnetic power especially useful because there would be no props or rudder to damage when the ship crunched through pack ice. The ship would also be fairly easy to control—to turn right or left, the skipper would simply reduce the current to the electrodes on one side while increasing it on the other. "The response," said Saji, "would be very rapid. All you need is a button."

Would the degree of salinity in seawater be a limiting factor?

Saji doesn't believe so. "We haven't researched the content of salt in the world's seas, but it doesn't differ all that much. Deep in the sea, however, salt density is increased, so electroconductivity also increases—and that's a plus for submersibles." According to Way, a magship could theoretically operate even in dirty river water full of organic material, but not as well.

What about a magship's efficiency?

Saji said the propulsion systems on his models are now about as efficient as those of a conventional ship. He maintained, however, that a real magship could be roughly 50 percent more efficient than a standard vessel. Thrust would be considerably enhanced. If his latest model, the *ST-4000B* icebreaker, is ever built as a full-sized ship, its displacement would be the same as that of Japan's conventionally powered icebreaker *Shirase*, but Saji estimated that its thrust would be about ten times greater.

The next step would indeed seem to be construction of some kind of large magship, but so far Japanese shipbuilders—though they talk about it in glossy R&D brochures—have not been all that impressed. One of the skeptics, a top officer of the Japan Marine Machinery Development Association, which investigates new methods of improving sea transportation, put it bluntly: "The magships don't have a chance of being realized commercially during the next century." Among his group's arguments is that existing technology is not yet able to produce suitable magnets for an economically feasible ship. The association warns of another pitfall: the ship's magnetic field might attract metallic debris or larger objects—even another ship. Others argue that the powerful generators required to pump electric current into the water might be too expensive.

Howard Stevens, head of the electrical division of the U.S. Navy's famed David Taylor Research Center in Annapolis, had similar reservations. "On the surface," he said, "the technology looks impressive. I feel, though, that it's one of those ideas that when you get down to the dollars and cents, the nuts and bolts, and the hard calculations about how much thrust you're going to get, the payoff just isn't there."

Saji had answers to most of the objections. Even before the new high-temperature superconductors came on the scene, he was predicting that superconducting magnets were sure to improve, eventually making a magship very economical. And while electricity for the electrodes is now provided by a conventional generator, in the future it may be possible to use a superconducting energy storage unit that would do away with the need for a standard generator at least for shorter voyages.

Saji did concede that the possibility of a magship's attracting metallic objects posed serious problems. But he proposed some solutions, among them magnetic shielding in the form of a special coating on the ship's bottom, for which he has applied for a patent. But the easiest way to prevent accidental magnetic attraction, he feels, is simply to keep the magships out of shallow water and away from other ships.

Saji envisions magships of the future as large vessels working vast expanses of ocean, or specialized vessels, like icebreakers, stationed in remote areas. Such ships would never come into a regular port but would tie up at artificial islands reached from shore by bridges and conventional ships. If a magship ever had to cross the path of other ships, some kind of special electronic signal (perhaps a SQUID detector) could alert them to its presence.

And so Saji sails on, buoyed every so often by some maritime technology expert who feels the physicist is on the right course. "When propellers were first introduced," he said, "they revolutionized the nature of ships. We think electromagnetic propulsion may well represent a similar advance over the propeller. People think ships like this are ten years in the future. They're not. If we had the money, we could build them today."

Mike Superczynski, head of the Machinery Technology Branch of the Navy's David Taylor Research Center, is a big man with a booming laugh who knows all about Japanese EMT but is far more interested in what the navy calls S/C electric drive—superconducting electric drive—as it applies to ship propulsion, and specifically to a test craft called *Jupiter II*. He also knows a lot about superconductivity and likes to tell the story about the admiral who once introduced him to a scientific symposium with the words, "Now the 'super' is Mike's middle name, and the 'czynski' is Polish for conductivity."

It is not, really, Superczynski was quick to explain, although it might well be, given the fact that the technology team of which he is a key part has, over the years, made enormous strides in developing both superconductive machinery for electric drives, and the sister technologies needed to make viable superconductive machinery systems. And now that the new superconductivity fever has hit, Superczynski and his people have been spending a good deal of time thinking about how the new materials might be applied to ship propulsion and auxiliary equipment, and whether or not high T_c would really improve reliability and reduce cost on the high seas.

The navy's approach to superconducting ships is far different from Saji's. While the *ST-500* bootstraps itself along using seawater as the power fluid, the craft the navy has christened *Jupiter II* runs on a superconductive motor which drives a propeller. Moreover, the *ST-500* is a model that has done its tricks only in a test tank. *Jupiter II*, with a 400-HP motor and generator (later upgraded to 3,000 HP), was the first vessel in history to be propelled at sea by a completely superconducting electric drive system.

The vessel's beginning—both the origins of its hull and the concept that drives it—was inauspicious. Before it went superconducting, it was a lowly workboat serving the offshore oil industry in the Gulf of Mexico. Reconfigured by the Coast Guard in 1976 and rechristened (she was named after the *Jupiter AC-3*, which pioneered the use of conventional electrical motors in 1913), she was no beauty. With her stubby hull and low, many-windowed superstructure, she looks more like one of those harbor cruise boats than the history-making vessel that has completed more than a dozen at-sea trials, the first to move advanced ship-drive technology out of the laboratory and into a realistic operating environment.

Jupiter II wasn't around during a visit to the navy's lab in Annapolis, but her engine was. Like Saji's model, it was up on blocks in a laboratory that looked, as do so many others where physicists and engineers work, like a machine shop full of jerry-built equipment stacked to the ceilings.

Superczynski points proudly at *Jupiter*'s drive system and, like a car salesman carried away, summarizes all the abstruse technical data. "Now this is it here," he says, giving the gleaming, cutaway hunk of machinery a pat, "3,000 horsepower, built by GE, and it's a superconducting homopolar drum-type machine. This is a superconducting magnet right here. These are the coils, the individual superconducting wires, niobium-titanium, in a copper matrix, three hundred strands in each wire, wound in solenoids, magnetic field toward each other. The net effect is a radial field out through this here region, and then out ultimately into the steel flux shield. All the superconducting magnet does is sit here and produce a radial field, coming out here, and the rest of the shield doesn't care where that field comes from, and the old magnet doesn't even care that the rest of the machine is out there. The rest is a cryostat, this stainless shell over here, the whole thing flooded with liquid helium, and the space in between is a vacuum, bearings at each end, and seals, and it all operates at 100 volts and 22,500 amps, so it's fairly high current, you can see."

Right! But in the language of the living, what's going on here?

Superczynski admits with a laugh that he didn't invent this sucker, that the patent is held by one Tim Doyle, who's no longer at the lab, and that, really, it all started back in 1847, with Michael Faraday, the Brit who discovered electromagnetic induction. Faraday's creation, a so-called disk-dynamo, was the simplest of all electrical machines. A copper disk was mounted in such a way that part of it, from the center to the edge, lay between the two poles of a horseshoe magnet. When the disk was rotated, a current was induced between the center of the disk and its rim by the magnetic field. By applying a voltage between the rim and the center of the disk, the disk could be made to rotate through the force produced by magnetic reaction. The disk, thus, could be made to operate as a motor of sorts.

"This thing here is a so-called homopolar motor," Superczynski says, pointing at *Jupiter's* insides. "The idea was never used, hardly, in any application because of an inherent problem: a very high current and low voltage, which makes the transmission of electrical power from one machine to another difficult. People kind of said, Well, we don't need that kind of headache, so let's go to the sort of machines that are high voltage and low current, which we can handle. But now comes along superconductivity, and its adaptation to this Faraday thing is very, very simple.

"In fact, this design of ours is the simplest application of superconductivity in a machine since it's the easiest place to put superconductors."

Essentially, the *Jupiter* propulsion system consists of a gas-driven electric generator that supplies power to a superconductive homopolar motor attached to the propeller. For all intents and purposes it appears to function much like a conventional motor rig—except for the superconducting elements and the refrigeration system. "When we talk about a superconducting electric-propulsion system for a ship," said Superczynski, "there's often a misconception that because it's superconducting you no longer need fuel in the tank. That's not true. You've got to have fuel oil, which fires either a steam or a gas turbine and creates a rotating shaft, and then when you have that rotating mechanical energy, you change it into electrical energy by a superconducting generator instead of a conventional one.

"It's really a transmission system rather than a total propulsion concept, like where you have a turbine that turns a shaft which turns a generator which is electrically connected to a motor which eventually turns the prop. That part's all the same, but the motors and the generators are intrinsically different, they're more power dense and more efficient."

ELECTRIC DRIVE
LONGER-RANGE EXPLORATORY DEVELOPMENTS
MOTOR COMPARISON
40,000 hp, 180 rpm

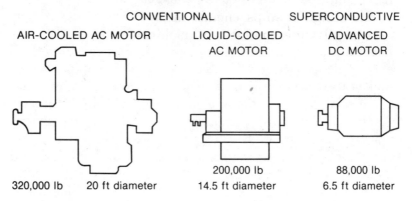

CONVENTIONAL		SUPERCONDUCTIVE
AIR-COOLED AC MOTOR	LIQUID-COOLED AC MOTOR	ADVANCED DC MOTOR
320,000 lb 20 ft diameter	200,000 lb 14.5 ft diameter	88,000 lb 6.5 ft diameter

Superconducting elements, as this comparison of motors demonstrates, make for a smaller and lighter shipboard engine.

It is the superconducting part that also makes the difference in size and weight of the engine. A conventional air-cooled AC ship-propulsion engine developing 40,000 horsepower at 180 revolutions per minute might weigh 320,000 pounds and be around 20 feet in diameter. A liquid-cooled AC motor of the same horsepower would be 14.5 feet in diameter and weigh 200,000 pounds. A superconductive advanced DC motor would have a 6.5-foot diameter and weigh 88,000 pounds. Aside from the smaller size and lower weight, an electric-drive system would be far easier to control (so, too, would be the speed and maneuverability of the ship), have much higher efficiency, run much quieter, and provide substantial reductions in overall size and cost of the ship.

Superczynski talks readily about the performance advances of S/C electric drive—even with the present liquid helium cooling system—over conventional shipboard systems. Still, though excited about the new superconductivity, he remains realistic, like most of the scientists working in the field. Prepared to wait and see, he puts the breakthroughs in the same category as nuclear fusion—great hope, but snagged by a torrent of engineering problems.

He discusses the advantages of an S/C system and what might be expected from the higher-T_c materials, including the effects a room-temperature superconductor would have on the design of ships' engines in the future:

> In all superconducting machinery, you really have one different thing—and that is that all of these machines consist of magnetic fields that are generated by turns of wire, which is essential to making a motor really work. The difference between a normal machine and a superconducting

SUPERCONDUCTIVE ELECTRIC PROPULSION
ESSENTIAL TECHNOLOGY

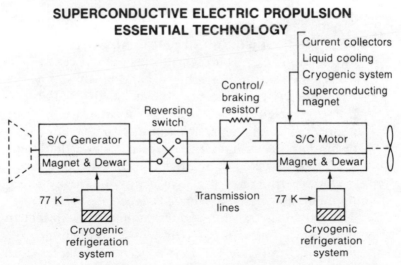

Essential technology of the Navy's Superconductive Electric Propulsion system for the Jupiter II. (*Courtesy of the U.S. Navy's David Taylor Research Center.*)

one, basically, is that the wire that creates the field is now a superconducting one, and this does several things for you.

You can generate high magnetic fields because the wire that is carrying the current does so without any electrical loss, and you can pack a lot of current into a very small part of the machine. As a result, you generate a flux, or an intense field level, which may be three to five times as intense as in a conventional machine. So you get higher power density in a smaller space, and because you don't have the field losses, you get improved efficiency.

You can also do things because of this capacity to carry large currents with no loss. You can, for example, get rid of the magnetic circuit that's in the machine—get rid of the iron, the thing that carries the flux. The only reason a machine works with any efficiency with normal materials is because there are iron magnetic circuits involved. With superconductivity, you don't have all that. There is some iron in these machines to shield the high fields from the outside world and to shape the field, to concentrate it where you like to have it go. But these things are not really necessary for the machine to operate, unlike a conventional machine where you need all that iron in the right place. We could get rid of it all, I suppose, but some guy walks by with a wrench in his hand, you have this large magnetic field, and he gets sucked right into it if he doesn't let go.

Now, getting rid of the iron has all sorts of implications. Because you've reduced the weight and size of the machine, you can place it remote to the ship. You can put it into a pod, reduce the drag, and improve the ship's performance, increase its range, do all sorts of things if it's smaller.

Of course, you have to keep all of this superconducting stuff cold. So you encase it all in a glorified thermos bottle, shoot in your helium, and let it go. Now, that's an added complication, since all this wire can't just sit out in the open. Right now, our cold box is a cubic meter, and the compressors about the same size. If we went to liquid nitrogen, it'd be easier to handle, but you still need a good insulator, still need containment, a closed system, a fridge. Only thing is, it operates at 77 K instead of 4.

DTNSRDC ELECTRIC DRIVE TEST CRAFT JUPITER II

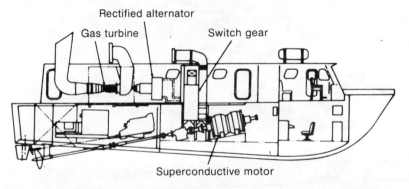

A cutaway view of the *Jupiter II* showing the components of a propulsion system which uses a superconducting DC motor.(*Drawing courtesy U. S. Navy's David Taylor Research Center.*)

Still, let's look at the advantages of a higher-T_c drive system, if we can get one to operate as well as the one we have now. The machine design itself would change only slightly, but the auxiliary systems would change dramatically. The refridge system would be different. First of all, when you need nitrogen, you can get it out of the air anywhere on earth. You need helium, you got to go to the gasworks in Kansas. Keep in mind that the fridge is going to look like the one that sits in your home—and the question I ask you is: When's the last time you needed freon for your fridge? This is a closed system, and what goes in for coolant stays in the coils. Indefinitely. If any of it boils off, you simply recondense it, put it back into your system. You don't have to replace it—unless you get a break in a line—and if you do, all you need to carry is a bottle or two. It's not like running out of gas in your car.

SUPERCONDUCTING MOTOR

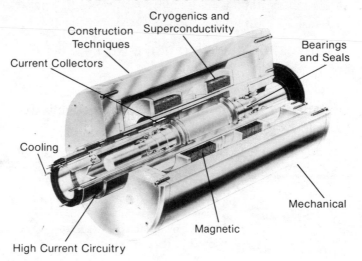

Cutaway of *Jupiter II*'s superconducting drive system. (*Photo courtesy U. S. Navy's David Taylor Research Center.*)

Now, if you're refrigerating at 77 K, that takes less electricity. Maybe a factor of ten lower. I have to say, though, that the power requirement, even at 4 K, is so low that it probably won't make all that much difference, in absolute terms, of course.

What people fail to realize amid all this interest in nitrogen temperatures is that most of the benefits of superconductivity come at any temperature. There's marginal improvement from 4 K to 77 K—it's not all that great. You'd get an inch reduction in the diameter of the insulation system needed to contain the nitrogen, that's all. Okay, now suppose room-temperature stuff comes along. Essentially, you'd eliminate your refrigerator, but you wouldn't change the engine design because the machine doesn't care what makes the flux. All I'm saying is that if you solve the problems for a helium system, you solve them for a nitrogen system, or a room-temperature system.

You know, there's a lot of work going on in high-temperature superconductivity, but there's no real push to make a refrigerator as reasonable as the one in your home. The only thing that stands between superconductivity and its widespread application at any temperature is that the refrigerator doesn't exist. If you had a cooler as reliable as the one at home, superconductivity would be everywhere.

It most certainly would be everywhere on a navy ship of the future. Superczynski uses an artist's rendition to show what such a ship might look like. In addition to superconducting propulsion machinery, there are these shipboard applications: an aircraft catapult, magnetic minesweep, high-speed computers, high-power radar, energy storage, electric weapons, and high-power sonar.

"Superconductivity works," said Superczynski. "No doubts about it."

10
FLYING TRAINS

THE FIRST PRACTICAL locomotive, a steam-driven behemoth designed in 1829 by the British engineer George Stephenson, managed to huff along with a full load of passengers at a speed of 24 miles per hour. Not bad for its time.

Railroad steam has gone the way of the paddlewheel, replaced on the nation's commuter and cross-country railways by diesel oil and electricity. For decades, railways themselves have been in decline. Intercity passenger travel is dominated by the automobile and the plane. The truck is king of the road for freight hauling. Speed and convenience are the obsessions that have forced the train off to a siding, if not off the track altogether.

Yet, recently there has been a renewed interest in rail alternatives—high-speed ones, of course, in keeping with the national preference. More specifically, the focus has centered on fast trains that would take advantage of developments in superconductivity and give us one of the most spectacular of all of its potential applications: magnetically levitated trains, maglevs, that would literally fly between our cities on electromagnetic cushions at speeds up to and beyond 300 miles per hour. "Both the air and highway modes have required massive investments in public and private expenditures," said Larry Johnson of Argonne Laboratory's Center for Transportation Research. "The question now is whether a form of high-speed rail is a legitimate option in the alternatives to satisfy future intercity travel demands in the U.S., and particularly if magnetically levitated trains . . . prove to be the preferred choice among high-speed rail alternatives."

As Johnson sees it, current interest in high-speed rail alternatives is stimulated by two factors. The first is increasing air traffic congestion, a problem not easily resolved because of the frequent controversies swirling around continued airport expansion and new airport construction. The last one built was Dallas–Fort Worth, in 1974. Second is the energy problem foreseen for the next decade, despite the current availability of oil in the world market. In the United States, petroleum consumption for transportation alone exceeds domestic oil production, a situation that is certain to get worse as oil reserves are further depleted and transportation demand continues to grow. Johnson believes that electrically powered, high-speed ground transportation addresses both of these issues.

Electricity is, of course, no stranger to the railbeds. Streetcars, subways, elevated systems, and commuter lines, more than five thousand miles of them in the United States, are familiar to most of us. The mechanism that drives them all is fairly simple: electrical current from a powerhouse is passed along to traction motors that drive the car or its wheels. The power can get to the trains in a couple of ways. Current may be drawn from overhead wires by a metallic collector-bar that slides along the wires and sends it down a conductor frame to the motors near the wheels. There is also a system by which electricity from a third rail alongside the track flows into the motors through metal shoes that slide along the rail. Large locomotives run on electricity, including the diesels whose engines drive a generator that produces electricity which, in turn, runs the electric motors tied to the train's wheels. The familiar Metroliners, which run between Washington, D.C., and New York City are electric trains that can attain speeds of about 100 to 120 miles per hour.

The most-advanced trains in the world can do better than that. The French Train Grande Vitesse (TGV) can get up to 170 mph while averaging 130 between Paris and Lyon. The electrically powered Japanese "bullet trains," the renowned Shinkansen, have a design speed of 160 mph, though they generally operate at around 130, averaging 100 mph over the 600-mile run between Tokyo, Osaka, and Hakata. British Railways operates diesel-powered high-speed trains on a number of long-haul routes at top speeds of 125 mph and average speeds between cities frequently above 90. (To give steam its due, it should be noted that in 1893 what was believed to be the finest engine ever built, the 999, roared between Batavia and Buffalo on a speed run, hitting 112.5 mph, the fastest man had ever moved.)

But even these fast trains have limitations. Chief among them is that they use conventional steel wheels against steel rails. The 170 mph that the French TGV can hit is probably the limit on such a conventional system. Larry Johnson explained it:

> At speeds of about 120 miles per hour and greater, there is significant wear on both wheels and rails. Federal Railroad Administration standards for conventional freight trains and commuter trains allow relatively large discrepancies between the level of one rail and another—1.25 inches for 80 mph operations. However, the FRA standard drops to 0.5 inches [sic] for 120 mph operations. By comparison, the French TGV standard is 0.16 for the 170 mph portions of the system. Satisfying the standards for high-speed service is not impossible, but it is very expensive.
>
> Second, the speed limit of conventional trains driven by wheel traction is limited by the fric-

tional forces that develop between the wheels and the rails. The use of linear propulsion motors, in which the wheels would be used only for suspension, would make maximum speeds of 180 mph technically feasible. However, the high maintenance costs associated with the required track alignment, smoothness, and curvature specifications may render it economically unattractive.

Third, the serious problems associated with noise and vibration of steel-wheeled trains have been major factors influencing the Japanese to examine other alternatives for intercity transportation.

The Japanese alternative is the maglev system, and several prototype trains have been tested successfully by the country's engineers. What is maglev, however, and how does it work? And how will the new superconducting materials contribute to giving us a transportation system that would be a realistic version of the fabled flying carpet?

There are actually two systems that employ magnetic force to move trains. One, in the Japanese design, harnesses the repulsive forces of superconducting magnets; this is known as the EDS system, for electrodynamic suspension. The other, as tested in Germany, uses the attractive forces of conventional electromagnets; it is referred to as an EMS system, for electromagnetic suspension. Each system has advantages, and despite their differences, both show great promise for ground transportation in the not too distant future.

The idea of flying trains did not, however, originate with the Japanese or the Germans. It was worked out on paper some twenty years ago by James Powell and Gordon Danby of Brookhaven National Laboratory, and Henry Kolm of the Massachusetts Institute of Technology. Stuck in a traffic jam near the Bronx-Whitestone Bridge in New York, Powell, then a twenty-eight-year-old nuclear engineer, thought how nice it would be to be able to fly over it all. Magnetism and levitation, he eventually reasoned, would be the way to do it. He got together with Danby, and the two published the first paper on a superconducting maglev in 1966. Two years later they had the first patent for the system. By 1974, at the Bitter National Magnet Laboratory, Kolm, supported by a small grant from the National Science Foundation and money from private sources, had built a 1:25 scale model of a maglev system. He called it a magneplane, and it moved—scant millimeters above a 400-foot aluminum track—at an astonishing 56 miles per hour.

Fascinating it was, but few people paid attention—except for the Japanese. They had read Powell and Danby's paper with the same keen interest with which years later, they would read the obscure journal detailing the discovery by IBM scientists of a ceramic compound superconducting at a record-high Kelvin. In 1970, they exhibited a model of the train at the Osaka World's Fair. By 1979, they tested another at speeds that hit a top of 321 miles per hour, a world record. In 1985, another maglev carried more than a half-million passengers on short runs at the science fair at Tsukuba.

The Japanese maglev system with its superconducting magnets relies heavily on the Brookhaven and Bitter Labs' concept and on the fact that like magnetic poles repel one another.

Prototype maglevs have been built by Japan Railways and by Japan Air Lines, but the most interesting is a streamlined 72-foot, 17-ton test vehicle that operates regularly over a 4-mile "track" 30 miles north of Miyazaki, on Kyushu, Japan's southern island. Operated from a control room at one end of the roadbed—there is no engineer aboard—it carries forty-four passengers and can reach 260 miles per hour. (It was an unmanned version that hit the 321-mph world speed record in 1979.)

The magnetic engineering system that drives the train is complex, but the principle behind it is easily understood, relying as it does on both the repulsive and attractive forces of magnets. The train is cradled in a U-shaped aluminum guideway, a sort of trough, which has thousands of electromagnets for levitation set in its floor along the length of the guideway; more electromagnets, for propulsion, are built in precisely arranged intervals into the inside walls of the guideway. Eight 5-foot, lightweight superconducting electromagnets, coils of niobium-titanium wire that are magnetized when electricity is pumped into them, are arranged along the sides of the train's undercarriage and cooled with liquid helium in cryostats, the thermos jugs that keep the coolant from boiling off.

To get it all rolling—indeed, the magtrain starts out that way, gliding down the guideway on support wheels—the engineer in the control house sends alternating current to the propulsion magnets set into the sides of the guideway. Because the current is alternating, the magnetic polarity of the propulsion magnets is continually reversed, enabling them to alternately pull and push on the train's magnets with a frequency that controls the speed of the train. When a train's magnet has an opposite polarity from one on the guideway wall, it is pulled forward; when it is the same, it is pushed. By manipulating the frequency of the alternating current, the engineer can control the pull and push, and hence the speed, in a smooth sequence as the train moves past each of the propulsion magnets. Because the superconducting magnets are so strong, and because the flow of electricity is without resistance, only low current is required for strong forward propulsion. The effect is like a surfer riding into shore on breaking waves. With the magtrain, the waves it rides in on are magnetic.

As the train picks up speed, the field from the superconducting magnets on the train meets the field from the electromagnets in the roadbed—they are of the same polarity, which means they resist one another—and the entire train lifts into the air. Suspended 4 inches off the tracks, without the friction of conventional rails to slow it down, the magtrain now is able to take full advantage of the alternating pull and push created by the repeated reversal of polarity, and it glides swiftly and noiselessly along on its magnetic cushion. The only time it uses its wheels is at liftoff and landing. Said Larry Johnson, "It's not a locomotive technology, really, but an aerospace one. What you have here is a 300-mile-per-hour levitated fuselage, and the principles behind that are those of an aircraft."

The German system, called the 06, is also levitated, but it does not depend on superconductivity for flight. It uses conventional iron-core electromagnets to provide a tiny (.5 inch) air gap between the vehicle and the track, and it is based not on magnetic repulsion, but on attraction. Developed for the Federal Republic of Germany by Transrapid International, a consortium of three West German multinational corporations, the 06 is a sleek monorail vehicle that has managed up to 250 miles per hour carrying two hundred passengers, over a 20-mile loop near Emsland. It has already logged more than 100,000 miles, and there is serious talk of installing a 185-mile line between Frankfurt and Düsseldorf. (The Japanese train, which is heavy and cumbersome, runs back and forth on its short, straight run, and does not have the ability to bank on a curve.)

The German train carries its magnets on the undercarriage, curving them around and under the crossbar of a T-shaped rail. Energizing the magnets pulls them up toward the underside of the crossbar, and the train's car is lifted into the air. Magnets in the track propel the train, as in the Japanese maglev. This system is simpler to build and less expensive to run but, thus far, is slower than the Japanese.

Apart from its speed advantage, a maglev system of either type is far superior to conventional trains for many reasons. Larry Johnson ticked them off:

> Non-contacting operation [that is, no contact between rails and wheels] means that rain, snow and ice will not pose significant problems for safe and timely operation. Because of the non-contacting suspension, maintenance costs associated with maglev transportation will be considerably less. There is virtually no mechanical wear and tear on either the track or the suspension system. For example, the steel-wheeled Japa-

nese bullet trains, running at 120 miles per hour, must have their tracks realigned each night. The lack of moving parts for a maglev train should greatly increase the dependability and reliability of the system. Being electrical, maglevs would not be dependent on dwindling U.S. petroleum supplies and the electrical energy can be provided by hydro, coal, or nuclear power. Further, the energy intensity of a maglev train would be on the order of one-fourth that of intercity aircraft or automobile travel on a passenger-mile basis. Because of the combination of higher achievable speeds, which increases the ridership potential, and the greatly reduced maintenance costs of a non-contracting system, a maglev may have the greatest potential among the intercity train options to operate on a revenue-sustaining basis. And last, noise and vibration, which reportedly is a major concern of the Japanese bullet trains, should be considerably less.

While either an attractive or a repulsive (magnetically speaking, of course) maglev system has more to offer than a conventional rail system, there is general agreement that the attractive EMS system, which uses the developed technology of electromagnets, appears to have the commercial edge at the moment. Speed might be sacrificed, but the initial capital costs of the primary suspension system would be lower with EMS than with EDS.

The advent of the new low-temperature superconductors, however, has whetted a lot of appetites for the repulsive EDS system, the one that uses superconducting magnets. This system seems to have advantages that could more than offset its higher start-up costs—if and when magnets of the new oxides come into practical use. Larry Johnson listed the benefits:

• Ceramic materials cost less, currently at $10 a pound versus $70 a pound for niobium-titanium.

- The high-T_c feature would permit the use of low-cost liquid nitrogen refrigerant, a savings on the order of twenty to one over liquid helium.
- The total refrigeration system could be simplified, with an attendant reduction in the amount of insulation required around the magnets; this would increase the flexibility, and further reduce the costs, of the design.
- The EDS levitation system is dynamically stable and requires no feedback controls to maintain clearances. The EMS system, on the other hand, is inherently unstable, necessitating gap sensors and accelerometers to regulate the power to the electromagnets to maintain stability.
- High-T_c superconducting magnets make higher magnetic fields possible, resulting in increased track clearances, which in turn relaxes design requirements. Because of the larger clearances, the system is more "forgiving" of discrepancies in rail level and problems associated with inclement weather. Furthermore, only superconductors provide the levitation forces that create large track clearances at reasonable cost. Stronger magnets also mean that less aluminum could be used in the guideways that serve as "rails"—by far the most expensive part of the system.
- Lighter weight superconductors would reduce the weight of the primary suspension system (and in turn, vehicle weight) compared to an EMS design using iron-core electromagnets. The weight reduction should further reduce the size and weight of the propulsion system.

Johnson also commented on the opportunities for maglev development in the United States:

High-T_c superconducting material is not an enabling technology, but it is an enhancing one. For maglev trains, this could prove to be extremely advantageous. While Japan has the lead, for now, in EDS technology, the U.S. can use its lead in high-T_c superconductors to leapfrog the current EDS technology with a more reliable, less expensive system. Although the U.S. is behind in maglev test facilities, a conventional maglev prototype system could be constructed in this country, drawing on the large body of work conducted in the U.S. during the 1970s, and updating it with the more recent advances in other countries. The development of a maglev system using conventional superconducting magnets could take place simultaneously with the development of commercial quantities of high-T_c superconductors that would meet the specifications of an advanced maglev system.

Besides the Japanese and the Germans, maglev research programs are underway in Canada, England, and Romania, and a dozen U.S. states are studying high-speed trains, both magnetic and conventional electric, for intercity passenger travel that covers distances between 100 and 600 miles. Trips less than 100 miles would probably still be dominated by automobile travel, though maglev backers feel there could be a market for some business travel. Trips beyond 600 miles would still rely on the airplane.

But for all its distinct advantages, a magtrain is still rather a hard sell in the United States even though the idea originated here and the technology to put a maglev system together had been available twenty years ago. The Department of Transportation looked into it during the Nixon administration, awarding grants to Ford Aerospace and the Stanford Research Institute. Some $32 million was even spent on a high-speed-vehicle test track at Pueblo, Colorado. But DOT eventually derailed the whole concept, citing the cost of building and laying new tracks, the difficulty of obtaining rights of way, and various complexities of such a system.

Maglevers like to tell the story, perhaps apocryphal, of

one pioneer in the field who contacted the department to say he had a revolutionary idea for transportation and would like to come in and talk. He hadn't mentioned a magtrain in his initial conversation and was asked to drop in, by all means. When he got down to specifics, a DOT official allegedly told him, "Oh, but we aren't interested in *ground* transportation!"

About all one can get officially today from government transportation officials is that maglev is interesting and intriguing but probably not a priority of the foreseeable future. This is probably why one hears comments from the Japanese like that of Masanori Ozeki, president of Japan's Railway Technical Research Institute. Asked why he thought the United States had dropped out of the maglev race, he replied, "I think the reason must be in undertaking civilian research and development: the U.S. has forgotten how to identify what is vitally important, and to be prepared in sharing the risk of developing it."

Still, the system has had its champions, among them New York's Senator Daniel Patrick Moynihan, who quietly introduced a bill in 1987 pushing maglev trains. He called his proposed legislation the FAST Act, for Federal Advanced Superconducting Transportation Act, a misnomer of an acronym to say the least, given the lack of interest and action that has attended it in the Commerce Committee, where it was still sleeping at this writing. Undaunted, Moynihan introduced another bill in 1988, carefully designing it so it would wind up in the Subcommitte on Water Resources, Transportation, and Infrastructure, part of the Environment and Public Works Committee, a panel which he just happens to chair.

Moynihan's argument was that the Japanese and German projects did not just happen. In a newsletter to constituents, he wrote:

> The Japanese government has poured about $1 billion into maglev research; the West German government, an estimated $700 million. Bonn plans to spend another $320 million over the next

10 years. By comparison, between 1966 and 1975, the U.S. spent $3 million on maglev. Then stopped. Even with this derisive investment, the Office of Technology Assessment in 1983 stated that "U.S. maglev research was [then] on a par with similar research programs at the time the U.S. government canceled it in the mid-70s." More than a decade later, we are well behind Japan and Germany—and perhaps even Romania. Yes, Romania!

The need for a maglev system in the United States, Moynihan declared, was quite clear.

The Interstate Highway System is just about finished—after 30 years—and cannot be expanded. There are limits to how many cars can use a particular corridor. By 2020 it will take 44 lanes to carry the traffic on I-95 from Miami to Fort Lauderdale. Pretty soon there won't be anything left of Florida! Just the other day, the new head of the Federal Highway Administration gave me their forecasts for the year 2000. In every urban area over 1 million population, traffic conditions will have declined to service level E or F—stop and go, or just plain stopped.

We have a means. The Interstate Highway System provides a perfect right of way for a national maglev system. It is already there. Paid for. Owned by the Federal Government. Surely some partnership can be worked out. Florida is interested in a Tampa-Orlando-Miami run. Nevada is exploring an interstate maglev train from Las Vegas to Los Angeles. I would settle for Manhattan to Riverhead along the route of the Long Island Expressway, 15 to 20 minutes.

The right-of-way issue, always a stumbling block when a maglev system is discussed, was also addressed by Gordon

Danby and James Powell in 1988, during testimony before Moynihan's committee. Powell noted that the system should rely to a great extent on existing rights of way, particularly those along highways, since new rights of way would greatly increase the cost. Danby offered another practical reason.

> A casual observer can see that much of our inter-state highway system is straighter and more level than would be required if control of personal discomfort due to forces on the body were the sole criterion of design. It follows that in many locations much greater speeds than are practical for autos could be used before passengers would become uncomfortable from the roller coaster effect of going over hills and around corners. Different speeds are possible. Starting and stopping already require variable frequency propulsion currents. In principle some locations could operate at different speeds from others to conform to the realities of available road beds to even length of trip.

Consider, as did Powell before the Moynihan committee, the following hypothetical maglev system built between two cities, 250 miles apart. With a top speed of 300 mph, trip time would be approximately one hour, allowing for pickups and dropoffs at several points in each city. At peak periods, several vehicles would be coupled together, with a maximum of a hundred vehicles in operation. The two-way track between the two cities could easily carry a hundred thousand passengers a day. Peak system operating power for air and magnetic drag, as Powell saw it, would be only about 150 megawatts. The energy consumption for the trip would be about 15 electrical kilowatt-hours per passenger, equivalent to the energy available from a single gallon of gasoline—extremely small compared to that of auto or air travel for the same 250-mile trip.

An additional 5 kilowatt-hours per trip would be needed to refrigerate the superconducting magnets on the levitated vehicles, a requirement that assumes that the present liquid helium system is used. High-temperature superconductors would significantly reduce the energy consumption. Since most of the modest electrical energy consumption would come from power plants, oil fuel use per passenger trip would be essentially zero.

"If all of U.S. transport were by magnetic levitation," Powell said, "approximately 2 billion barrels of oil would be saved a year, resulting in greater energy security and much more favorable balance of trade. It is, of course, unrealistic that magnetic levitation would completely replace all other forms of transportation, but it could make a very significant contribution in reducing oil use."

And the cost? Powell told the committee:

> We estimate that the two-way track cost will be in the range of $2 to $3 million per mile, not including right of way or grading as required. This is based on compact, prefabricated track structures that can be installed at grade or on preset pylons as needed with a minimum of field construction. The 250-mile track would cost approximately three-quarters of a billion dollars. With an additional one billion for stations, vehicles, and power, the total investment would be on the order of two billion dollars.
>
> Although this system has extremely high passenger capacity, initial traffic will probably start low and build with time. We assume an average of 30,000 passengers per day for the first few years, at an average fare of $30. This is roughly comparable to auto travel and about half that of air. We also assume that half of the fare covers operating cost—personnel, maintenance, power, and advertising—and the rest retires the investment. Under these assumptions, investment payback is 10 years. This is probably acceptable, since traffic increases will reduce payback time. This example ignores freight revenue during off-peak hours.

But will maglev ever actually get off the ground in the

United States?

The skeptics say no, that the projected construction costs are far too low, that the new superconductors are unlikely to cut costs all that much since current designs allocate only 1 percent of capital cost to levitating magnets, that the technology is no way near being ready to apply in such a radical way. The proposed Los Angeles–Nevada maglev line presents another special problem. Southern California supplies some 40 percent of the millions of gamblers who visit Las Vegas every year, and cheaper, faster train service would be expected to increase that number by another 25 percent. But while maglev might be heaven-sent for the gaming tables, it could mean a traffic and parking nightmare on both ends of the line as hordes of cars drive to and from the train. California legislators were also not all that thrilled over a plan that would send California money to another state.

Yet, the feasibility studies go on, perhaps spurred by the threat of U.S. technology exploited, once again, by foreign countries, notably Germany and Japan, which already are involved in planning for short demonstration maglev lines in Las Vegas. As Senator Moynihan put it:

> It is little wonder that the first commercial maglev transportation system built in the United States, in Las Vegas, will not be built by an American firm. The contract has been awarded to a West German company. Work has already begun, and will be finished by 1990. It is more a tourist attraction than a transportation system. But there it will be. And when you get on that vehicle, it will say: Invented by an American Scientist. Made in West Germany. And if anyone would like to watch a video cassette of the Japanese maglev demonstration model at the Miyazaki test track, write and I will pass on the request. You can watch the video tape—invented in the U.S. in 1956—on your VCR. Your VCR, made only in Japan.

11

IS THERE A SUPERCONDUCTING CAR IN YOUR FUTURE?

NEARLY TWO-THIRDS of all electrical power produced is used in motors, and their efficiency ranges from 75 to 95 percent. High-temperature superconductors may one day increase that efficiency to 99 percent, while reducing by as much as 25 percent the cost of the larger ones that produce 1,000 horsepower or more. There are two reasons why those large motors would benefit most from superconductivity: one is that they use substantial amounts of electricity, so small increases in efficiency can still yield substantial cost savings; the other is that compared to small motors, the large superconducting ones would need less refrigeration per horsepower of output.

Superconducting engines and magnetic systems using the old superconductors are already in use to drive ships and trains, but until very recently no one had put one of the new ceramic superconductors into a working electrical motor. That was accomplished in 1988 at Argonne. Called the Meissner motor because its operating principle is based on the Meissner effect, that property of superconductors that causes them to expel lines of magnetic force that can repel a magnet placed nearby, the Argonne motor consists of an 8.5-inch, circular aluminum plate with twenty-four small electromagnets mounted along the bottom of the outer edge, and two hockey-puck–shaped disks of yttrium–barium–copper oxide.

To run the motor, the ceramic disks are first brought to a superconducting state by cooling them to −290° F with liquid nitrogen. When the electromagnets are switched on, the superconductors respond by producing their own magnetic field, which pushes the magnets away. The aluminum plate, which is set above the two disks, rotates as its electromagnets spin past them. When the liquid nitrogen evaporates and the ceramic warms up, the motor stops.

The experimental motor managed only 50 revolutions per minute (insignificant when compared to the 3,600 rpm of the superconducting generator's spinning rotor), but it worked, and that was all that its inventors wanted. "We built it to show that simple, operating motors can be made with new superconducting ceramics," said Argonne's Roger Poeppel. "It's too small for practical use and produces negligible power, but it does demonstrate for the first time that these things are possible."

A 3-kilowatt, 140,000-rpm superconductor motor using the same ceramic is now being developed by a research team at Allied Signal, Inc., of Los Angeles.

Meanwhile, at Cornell University, engineers have produced a high-speed, virtually frictionless bearing that also depends on the levitating effect of the new superconducting ceramics. Bearings, whether ball bearings, roller bearings, journal bearings, or hydrodynamic bearings, are used to reduce friction in all machines in which one part turns or slides on another. There are also high-speed magnetic bearings in use today, swift-spinning devices that can achieve speeds of more than 100,000 rpm in a vacuum. Essential to a wide range of technologies, they often require complicated control circuitry to maintain their stability. The new superconducting bearing is expected to overcome that complexity simply because the levitating effect of the superconductor is self-stabilizing; that is, the forces that make it work also hold the device in place.

The unique bearing is the brainchild of Francis C. Moon and Rishi Raj. After precisely measuring the magnetic forces generated by the yttrium superconductor, they made a small chunk of it using a sol-gel process, in which fine-grained ceramic powder was suspended in a solution, then heated to form a solid mass. They machined it to the proper bearing shape, cooled it to superconductivity with liquid nitrogen, and got it to levitate a rotor containing rare-earth permanent magnets. The bearing whizzed around at 66,000 rpm, respectable enough but with the potential of allowing rotors to turn several times faster than any bearings now in use—300,000 to 1,000,000 rpm in a vacuum. Such phenomenal speeds would allow smaller rotors with lower friction and less wear and heating problems to be used in gyroscopes, those spinning disks which are used, among other things, to maintain the stability of ships by damping their rolling movements.

Can these superconducting motors and bearings be combined to create that nonpolluting, quiet, simple-starting dream vehicle, the electric car? Perhaps. Room-temperature superconductors seem to be close at hand, and engineers have been able to build smaller motors and chassis with lighter materials, and space-age fuel cells that are similar to batteries but which draw their energy from a chemical fuel that is supplied while the cell is in use.

The concept of an electric car is by no means new. The first one showed up in 1834, long before gasoline-powered cars appeared in the United States and they soon caught on here and in Europe because of their low maintenance, few mechanical parts, dependability, and ease of operation. By the turn of the century, in fact, some three thousand of them were operating in this country. But though their speeds were good for their day (some eventually got up to 50 to 60 miles per hour), they were relatively heavy because of their lead-acid storage batteries and were expensive to operate. In addition their range was limited by the need to recharge their batteries every 50 to 80 miles. In 1911, the invention by Charles Franklin Kettering of the self-starter, which did away with the necessity of hand-cranking internal combustion automobiles, effectively put the electric car out of contention. Although electric traction managed to hang on in urban delivery vehicles, primarily in Europe, and in golf carts (the first patent for one of those ubiquitous, 10-mph vehicles was the so-called Arthritis Special, awarded to Texas oilman R. J. Jackson in 1948), Americans generally lost interest.

Not until the late 1960s, when concern over the environment became a heated issue, did interest revive. Gasoline cars were now a major source of pollution and were rapidly eating into world petroleum reserves. Everything old was new again, and electric cars suddenly looked like a good idea. General Motors quickly developed two experimental vehicles, the Electrovair and the Electrovan. The first was powered by silver-zinc batteries and had a top speed of 80 mph and a range of 40 miles. The van was run by an experimental power cell. By 1967 GM was building a series of small experimental electrics, as well as electric/gasoline hybrids. In 1973 the oil embargo showed the danger of dependency on foreign oil, and in 1988, Worldwatch Institute, a United Nations–sponsored group, issued the ominous warning that with some 400 million automobiles now in use, it was only a matter of time before "auto cultures" triggered a mammoth greenhouse effect and other monoxide-related health problems.

Will superconductivity raise the electric car to a level above that of a glorified golf cart? A superconducting electric motor, drawing energy from new lightweight batteries that would rarely need to be recharged, would be efficient enough to move a standard-size car over miles of highway at a good speed. It would, however, have to be small, something that might be difficult to achieve with the helium-cooled materials, since to keep cold even a small engine with niobium-titanium windings would require refrigeration comparable to what would go into a larger machine.

Robert Eaton, president and group executive in charge of GM's technical staffs, put the matter of the electric car in its proper perspective recently when he cautioned about being too optimistic about any near-term automotive application of the new superconductors. "No one," he said, "has yet been able to sustain superconductivity at room temperature, and that is the key to any commercial application in the auto industry. And even if a breakthrough were in sight, the use of superconductivity in the automobile would still take years to evolve."

Hopefully, given the ingenuity of American, Japanese, and German engineers—and the advances that are sure to come in superconductivity in the next few years—those rather pessimistic words will be retracted.

12

SUPERCONDUCTING TECHNOLOGY AND THE GENERATION, STORAGE, AND TRANSMISSION OF POWER

EFFICIENCY. THE WORD crops up virtually every time superconductivity is mentioned, and it is not difficult to see why. Efficiency, after all, is simply doing more with less effort, and when a superconductor carries electricity with no resistance and essentially no loss—and, theoretically, forever without any decrease in flow—that's efficiency of a most enviable kind. And although the transition from conventional superconductors cooled with liquid helium to the new high-temperature ceramics cooled with liquid nitrogen would not improve superconductivity—how, after all, can one make a perpetual motion machine more perpetual?—it would dramatically lower costs, simplify refrigeration systems, and improve the reliability of just about everything electric in which the ceramics are used.

A major beneficiary would be the energy industry. Utilities invariably suffer losses during the generation, transmission, and storage of electricity, and the possibility of sending current through a wire or cable over long distances with an efficiency approaching 100 percent would cut the United States' relative inefficiency in fueling its energy requirements. We now spend approximately 11 percent of our gross national product on energy, compared to 5 percent by Japan, a situation that costs us $220 billion a year, according to experts in the field. There would also be a plus for environmental health, given that the world's current energy systems are irrevocably altering our climate by spewing 5.4 billion tons of carbon into the atmosphere every year—more than a ton for each person on the planet. Let's look at how superconductivity might affect the big three of electric power technology: generation, storage, and transmission.

POWER GENERATION
Electrical Generators

Throughout our discussion of superconductivity, we have frequently likened the phenomenon to a perpetual motion machine because of the peculiar ability of a superconductor to allow current to flow indefinitely through it. In numerous tests, electricity has been shot through and around a superconducting loop for years, and scientists, using extremely sensitive instruments, have been unable to measure any drop in the current as small as 1 part in over 100,000,000,000.

The analogy to a perpetual motion machine is also apropos since the spiritual descendants of the inventors and dreamers who had tried for centuries to build such imaginative devices as self-turning water wheels and magnetically powered conveyor belts often tried to apply the concept to generate power to drive a motor. Such a notion might go something like this: If you attach the shaft of a motor to the shaft of a generator, the motor will drive the generator, which will, in turn, provide the power to run the motor. Nice try, but according to the law of conservation of energy, it won't work. Some energy in all of this turning of shafts would always be converted into heat by friction, and this is what ultimately stops all of the so-called perpetual motion machines—unless energy is supplied from somewhere to keep them going.

A superconducting wire, on the other hand, may be considered a perpetual motion—of the third kind, as the scientists say. Perpetual motion of the first kind, not feasible, is motion in which a mechanism, once begun, would continue indefinitely to perform useful work without being supplied with energy from outside. The second kind refers to motion in which a mechanism draws heat from a source and converts all of it into some other form of energy. Perpetual motion of the third kind is motion that continues indefinitely but without doing any useful work—such as the random motion of molecules in a substance, or frictionless bearings turning indefinitely in a vacuum. Insofar as superconductivity is concerned, it's perpetual motion of the third kind when electricity is flowing continuously around a loop—but only if you ignore the energy required to cool the wire to a superconducting temperature.

But no matter what the perpetual motion classification, high-temperature superconductivity, if applied to the generation of power, could cut the cost of electrical generators by as much as 60 percent, reduce operating costs of large electrical motors by as much as 25 percent, and give the United States the equivalent of around 15 percent additional generating capacity by allowing utilities to use existing generating facilities more efficiently.

To understand why, let's first examine generators, the ordinary, huge ones that are bolted to the floors of the electric power plants. Their design is fairly standard. Each one has a stationary, hollow, cylindrical iron core, a stator, wrapped in hundreds of windings of conducting copper wire. Within the cylinder is a wire-wrapped electromagnet, the rotor, which is fed direct current by a small, separate generator called an exciter. When an external source of mechanical energy is applied—it can come from fossil fuel, water that has been stored behind a dammed reservoir, or a nuclear reactor—it drives a turbine, an arrangement of wheels with fanlike blades mounted on a shaft, by pushing against the blades. As the shaft spins, it turns the generator's rotor, which creates a revolving electromagnetic field; this spinning field, in turn, produces a flow of current from the conductors in the stator, and it is this charge that is transmitted as electricity. In the case of power plants, alternating current is generated since it is easier to transmit than direct current. Typically, some 300 megawatts per generator are put out.

Such a generating system has been the sturdy backbone of the power industry ever since its emergence after 1870. But as efficient as it is—and a typical AC generator is 98.5 percent efficient in producing electricity, and 95 percent of that power reaches the consumer—things could be better. For one thing, the size of conventional generators has been fast approaching both technological and transportation limits. There are also the generator losses, relatively small, but still significant over the long haul and enough to set energy scientists and economists to thinking about the advantages of a superconducting system that would be smaller, potentially less expensive to build, and more inherently reliable. If the resistive power losses in the rotor could be removed by using a superconducting generator, efficiency could be boosted to 99.5 percent, compared to 98.5 for the conventional equipment.

The improvement may sound insignificant, but that 1 percent reduction works out to saving about $1.5 billion. Over the design life of a typical large generator rated at 1,000 megawatts—about forty years—this loss reduction would save the equivalent of more than four million barrels of oil. Moreover, if the new ceramic superconductors could ever be wound into a generator, there would be huge savings in both the energy and the equipment needed: refrigerating to 77° K takes only about 1 percent of the energy required to cool at 4° K, saving $250,000 a year, according to estimates by the Institute of Electrical and Electronics Engineers.

Among those who thought hard on the subject—indeed, they were the first into the field in a big way—were researchers at the Westinghouse Electric Corporation. Named for the American inventor and engineer, George Westinghouse, a pioneer in the introduction of alternating-current power in the United States (and whose first invention of note was the railway "frog," a device that allowed trains to cross from one track to another), the company's interest and involvement were in keeping with its long history of setting milestones in place in the field of power. In 1893 it produced the greatest display of incandescent lighting up to that time at the World's Columbian Exposition in Chicago. Westinghouse made the first industrial atom smasher in 1937, began generating power at the Grand Coulee Dam in 1941, developed the first atomic engine (for the first atomic submarine, the USS *Nautilus*) in 1953, produced the electricity for the first nuclear power plant in the United States in 1957, and built the country's most powerful water-wheel generator, for Niagara Falls, in 1960. In superconductivity, Westinghouse developed, in 1962, the first commercial superconductor, niobium-titanium, today's most important conventional superconducting material.

In 1973 and 1978, Westinghouse designed, built, and tested 5- and 10-mega-volt-ampere superconducting generators, the world's first multimegawatt superconducting machines. But the company's most ambitious project, announced in 1979, was a $19 million 300-mega-volt-ampere superconducting generator to be jointly funded by the Electric Power Research Institute of Palo Alto, California, the U.S. utility industry's major research arm. It was to be half as heavy (350,000 pounds) and more compact than conventional generators, would be used with a steam turbine in a fossil-fueled power plant, and would have a nominal efficiency of 99.4 percent. "We hope to start taking orders in the 1980s," exulted Eugene J. Cattabiani, executive vice president for power generation, "and we expect that as much as 50 percent of the new generating capacity installed in the U.S. power plants in 1995 will have this equipment."

The operating principle behind a superconducting generator is virtually identical to that of a conventional rotating generator, just as the superconducting ship-drive system we discussed is similar in concept to an ordinary shipboard propulsion system. Conductors in the spinning rotor

SUPERCONDUCTING GENERATOR

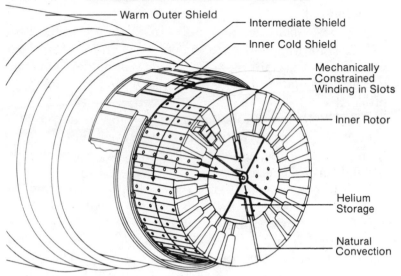

Warm Outer Shield

Intermediate Shield

Inner Cold Shield

Mechanically Constrained Winding in Slots

Inner Rotor

Helium Storage

Natural Convection

Cutaway view of the rotor of a superconducting generator. While the operating principle is the same as that of a conventional, rotating generator, the rotor is different. It is hollow and vacuum-enclosed, and instead of copper conductors, it contains a niobium-titanium superconducting field winding held in place by radial slots.

stimulate a revolving magnetic field that produces current from the surrounding stator. That is as basic as what goes on when you turn the ignition switch of an automobile and step on the accelerator. There are, however, some vital differences in both the rotor and the stator of the superconducting generator.

The most important change is in the rotor. In a superconducting generator, the rotor is hollow and vacuum enclosed; instead of the traditional copper conductor, it contains a niobium-titanium superconducting field winding held in place by radial slots around the rotor. With the conductors, the niobium-titanium wire, thus kept at or below 4° K, they conduct the current needed to produce the generator's rotating magnetic field with no energy loss. The centrifugal force of the rotor spinning at 3,600 rpm pumps the cooling liquid helium through the rotor, which is surrounded by a series of heat shields, which also serve to

keep the AC field from the stator out of the superconducting windings.

Any electrical current induces a magnetic field. That
field is also proportional to the current and to the number
of coil turnings. Coils made with superconducting wire can
carry higher currents and thus make for more compact,
efficient electromagnets: the relatively large field current
and small size of the superconducting rotor, in fact, creates
a rotating magnetic field about five times more intense
than that of a conventional rotor, and with a very low
expenditure of energy.

This intensity also affects the design of the stator, the
stationary component, that surrounds the rotor. Ordinarily, a stator with copper-wound coils needs an iron core
and laminated magnetic steel teeth to shape and focus the
rotor's magnetic field around the stator coils. The superconducting generator does away with all that heavy metal,
substituting a novel method of winding its coils—engineers
call it a spiral pancake—that makes for a mechanically
stronger generator. This design is more resistant to vibrations than conventional stator windings, with their complicated, convoluted twists and turns.

Apart from reducing the generator's weight and making
it more reliable, eliminating the iron allows the superconducting generator to produce far higher voltage because the
spiral design of the coil does not require the elaborate
grounding that limits the output of conventional generators. (Electrical supply systems are grounded to prevent
overvoltage and to improve performance.) Increasing the
voltage would mean that expensive step-up transformers—the devices that convert electricity from low
voltages into the high readings required by the transmission system—can be eliminated.

But despite such efficiency and the dollar savings for the
utilities, no superconducting generators are in service anywhere in the world. Several are under development here
and abroad—most notably in government-backed projects
in the Soviet Union, where a 300-megawatt machine is
being built, and in Japan, which has committed $200 mil-

lion to a ten-year program to develop a variety of machines. And what of Westinghouse's project, which was to have been installed at the Gallatin power plant of the Tennessee Valley Authority in 1986? It went down the tubes in 1983 after the Electric Power Research Institute, in a move that apparently surprised Westinghouse, canceled its participation in the project. The EPRI cited cost overruns, shifting research priorities, poor economic conditions in the power-generating sector of the utilities market, and a general declining interest in the technology.

Westinghouse and EPRI executives and researchers were, needless to say, disappointed. Some $20 million had already been spent, half by each partner, and at the time of the cancellation, Westinghouse had already completed design and engineering and was just embarking on the manufacturing phase. But despite the setback, Westinghouse stood by its faith in superconductivity as vital to long-term power generation requirements. In its statement, the company said:

> Results to date show our technology can be economical at utility ratings of today because superconducting generators are smaller, more efficient and more resistant to system disturbances than units being built with current technology. EPRI and its sponsoring utilities were satisfied with the design. EPRI requested that we preserve all useful design data, technology, and information developed to date. The industry has viewed the superconducting generator as a design for the next generation of electric power plants, those coming on line for the most part after 1990.

Westinghouse has not abandoned the work. Now that the new ceramic superconductors have arrived on the scene, the company is trying to work them into a 20-megawatt generator which the air force would like to use in a proposed high-speed, high-altitude plane powered by liquid hydrogen.

Hydrogen would be used as the propellant in the United States' National Aerospace Plane, which is scheduled for a maiden flight sometime in 1994. Known as the *Orient Express* because it could conceivably fly from Washington to Tokyo in two hours, the plane would travel through space on rocket engines, and within the atmosphere on hydrogen-fueled, air-breathing jets. The Soviet Union has already successfully tested a hydrogen-powered aircraft.

John Hulm is Westinghouse's chief scientist, a British-born physicist whose prime responsibility is to ferret out new technologies and keep open the lines of communication to the world's scientific community. Superconductivity is, however, his pet interest, and in the 1950s he and George Hardy, both then at the University of Chicago, discovered the first of the class-II high-temperature superconductors, the metallic alloys. In the 1960s, his researchers developed the world's first commercial superconducting wires and magnet coils.

His passion for superconductivity has been increased by the discovery of the new class of relatively warm superconductors, but he tends to be cautious when asked about practical applications.

> High-temperature superconductivity has set the world of theoretical physics on its ear. No one predicted it and no one knows where it will lead. Now that the mental barriers have been broken, each month brings new, optimistic findings that increase the promise of applications not yet imagined. There's so much work to be done yet, and we have to go through the same slow process that we went through when we were working with niobium-titanium. It's going to take years, and it would be tragic to delay indefinitely the building of superconducting generators, or the Superconducting Supercollider, while we wait for improved, warmer superconducting magnet coils that we might not ever be able to build.
>
> Even if the ceramics prove impractical, a su-

perconducting generator will come. It nearly broke my heart when the EPRI withdrew funding for the generator project. If we had not stopped development, we would now have a machine in a power station. That shutdown was not based on disenchantment with the project or on a lack of engineering skills. The bottom just fell out of the power market. The rise in the cost of fuel had driven up the cost of electricity, people were turning off their lights to conserve, and there simply weren't any machine orders.

I do think it'll be revived in the U.S. I feel we could start up again tomorrow and build those machines, without waiting for the ceramics. Niobium-titanium is good, but you also have niobium-tin, with the advantage that it'll withstand twice the field of niobium-titanium and you get pretty good current density.

Why don't we just go ahead? The ceramics still need plenty of inventive engineering to make them work. They're brittle, easy to crack. Compare that to a half-inch bar of copper that can be stretched into extremely thin wires without cracking. You have to be able to construct a decent conductor if you want to use the high-temperature superconductors in power applications—and by that I mean some form of massive conductor, not a plain strand of wire. It's got to carry thousands of amps of current per square centimeter at a high field, 5 to 10 tesla, and at 77° K. Well, we just aren't there yet.

Look at the Japanese. Their project is based on the old superconductors, and what they're saying is that should the new ceramics develop to the level of being used, they'll simply substitute them. They have a lot of confidence that they can do the engineering work with the old superconductors, as I do. They've studied what we did, ad nauseam. They've been out to visit, as they always do. And if they're lucky, the chances are they'll come up with something cooled at 77° K.

Fusion

For decades, physicists have dreamed of a safe, cheap, and limitless source of energy that would allow the world to thumb its nose at fossil fuels. All indications were that they had found it when they harnessed the energy locked in atomic nuclei and, in 1945, exploded the first atomic bomb. Nuclear power plants eventually went on line and proliferated. Their design was essentially the same as that of electricity-producing plants that burned fossil fuel. Only instead of coal and oil, the fuel was, and is, pellets of uranium, whose atomic nuclei are split apart by the same process that produced the atomic bomb, fission. As the nuclei of heavy, larger atoms burst into smaller fragments, enormous energy is released in the form of heat. The heat vaporizes water to create steam, which then propels the blades of a turbine, which in turn drives the generator that produces electricity. Marvelously, it doesn't take much fuel to do the job: an ounce of fissioned uranium-235 provides the same amount of energy as 388.4 barrels of oil.

That is the advantage of fission. Its drawback is the deadly radioactivity it generates, particles whose mass, from one type of reactor, is almost equal to the mass of the fuel consumed. Waste from a fission reactor typically requires thousands of years before it breaks down into biologically safe levels. Fission reactors are also relatively inefficient. They can use but a single isotope (atoms of an element that have the same number of protons but a different number of neutrons) of uranium, U-235, which makes up less than 1 percent of natural uranium ore. (More than 99 percent of natural uranium is nonfissionable U-238.) So-called fast breeder reactors might overcome the supply limitation by breeding fissionable fuel from U-238. But the fuel it produces from the uranium is plutonium, the same stuff that was inside the Nagasaki bomb—not an ideal by-product in a politically unstable world.

Ideally, the by-product fuel from a breeder reactor could

not be used to make a bomb. Such fuel would also be abundant and obtainable at low cost. It would have far less radioactivity associated with it. There would be less waste, and what there was would decay in a few hundred years, not thousands. Because lesser amounts of fuel would be required, the possibility of a nuclear explosion would be eliminated.

One nuclear reaction seems to fill the bill: fusion. The awesome process that powers the sun and the stars, and the hydrogen bomb, fusion occurs when the nuclei of smaller, lighter atoms are squeezed together—fused—under intense heat to form larger, heavier, and more stable nuclei. The reaction produces enormous bursts of energy which, if controlled and made self-sustaining, could provide a virtually unlimited supply of power from so innocuous and plentiful a source as seawater, and do it far more safely than the fission process.

FUSION REACTION

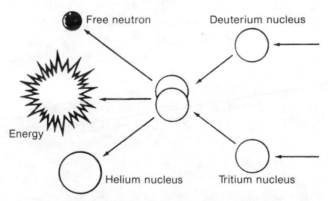

When deuterium and tritium fuse, helium is formed and a burst of tremendous energy is created. The energy produced by the hydrogen bomb, the sun, and the stars is the result of fusion reactions.

But controlling a fusion reaction is a formidable job. Extraordinary conditions are required for igniting the reaction and for containing it. Decades of effort by scientists in a number of countries to tame the process have thus far been unsuccessful. The hydrogen bomb, far more powerful than the A-bomb, is an example of a successful thermonuclear fusion, but it is an uncontrolled reaction, and one that depends, rather crudely, on the incredibly high temperatures of an exploding atomic bomb to start it up. For fusion's explosive force to be tamed and applied to more useful purposes—to supply electrical power, for example, or to study particle physics—reactors capable of withstanding heat several times that of the sun would have to be built, a far more difficult problem than loosing the massive power of the sun to destroy large cities in one gigantic, unbridled explosion.

The production of fusion energy would begin, of course, with fuel. In a fusion reactor, the most promising fuels are deuterium and tritium, the heavy isotopes of hydrogen, both of which can be extracted from seawater. Deuterium's potential as a plentiful energy source is easily understood when one considers that a small amount can produce the equivalent of some 300 gallons of gasoline when it is burned off.

The process would work like this: When the hydrogen atoms are superheated to temperatures measured in millions of degrees, their electrons are stripped away, producing a mix of free electrons and atomic nuclei, a hot gas of charged particles called plasma. Ordinarily, the separated nuclei would repel one another because of their positive charges. But if the heat is intense enough—and the heat that roars through a fusion reactor is five times that generated in the sun's core—the nuclei speed up tremendously and smash into one another with such force that they fuse together, form larger nuclei, and release enormous amounts of kinetic energy in the form of neutrons, energy begging to be captured and converted into electricity. It is as though a horde of microsuns, or miniature hydrogen bombs, has been unleashed—and tamed.

But all of these colossal events have to be confined somewhere, and a daunting problem has been to build a container so enormously strong that it can contain the superhot plasma, a seething mass of pure high temperature, perhaps on the order of 200,000,000° to 1,000,000,000° K. It reminds one of the classic conundrum of the so-called universal solvent, the theoretical liquid capable of dissolving any substance. What substance could be made into a vessel able to hold such a solvent?

It turns out that plasma can be contained—not in a material container, but in a magnetic "bottle," coils of wire that create magnetic fields powerful enough to confine the intense heat of the fusion reaction. Such magnets have already been built and tested for experimental use—with superconducting coils, which are preferable to copper. The reason? With copper, heat losses, which waste the energy needed to sustain the magnetic field, are too large to allow the construction of an efficient reactor; superconducting coils contain the heat and require a lot less energy to keep cool. Six large magnets—niobium-tin coils made in the United States and niobium-titanium coils made in West Germany, Japan, and Switzerland—were successfully tested recently at the Oak Ridge National Laboratory in Tennessee. The coils, 20 feet tall and weighing around 40 tons, each achieved peak magnetic fields of 9 teslas, or about 180,000 times the earth's magnetic field, after cooling with liquid helium. The force exerted on each coil was enormous: exceeding 5,000 tons.

Thus far, the most successful type of magnetic container is the so-called tokamak, a doughnut-shaped magnet first proposed (along with the physics of fusion) in 1951 by the youthful Andrei Sakharov, the Soviet nuclear physicist and peace activist, and Igor Tamm. Giant coils of superconducting niobium-titanium (or niobium-tin) trigger a pair of magnetic fields, one horizontal and one vertical, that force the charged plasma particles to circle around inside the doughnut without allowing them to hit the container's walls. Thus directed and confined, the density of the speeding particles is increased, and under the influence of the intense heat, they collide, fuse, and release their bursts of energy. Another ingenious confinement method is the magnetic mirror—in which the boiling plasma is trapped between two strong magnetic fields at the ends of a straight containment vessel.

Whether the new ceramic superconductors will ever find

TOKAMAK REACTOR

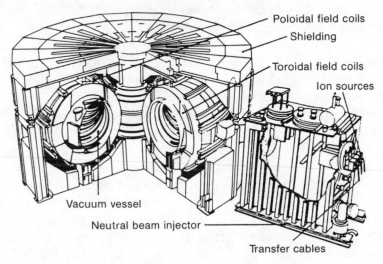

Major components of a tokamak fusion reactor are shown here. Giant coils of superconducting metal create powerful magnetic fields that confine the superhot plasma—a hot gas of charged particles—in a "magnetic bottle."

a place in such a man-made hell is debatable. Radiation from plasma changes the structure of many materials, making them brittle, and the superconductors would inevitably be exposed to some irradiation by neutrons. Since there is some evidence that these new materials are more sensitive to neutron damage than the conventional materials, their ability to withstand such radiation damage is of great interest. If they eventually prove to be resistant, the cost of shielding magnets would be greatly reduced, and fusion power would live up to its expectations.

Magnetohydrodynamics (MHD)

This is yet another energy source that could become economically feasible if the new high-temperature superconductors prove out. In converting various sources of energy into electricity, scientists have concentrated their research and development efforts on rotating types of equipment. MHD power, on the other hand, requires none of the moving parts typical of a conventional generator. It is generated by forcing ionized combustion gases—hot coal gases, say—through a powerful superconducting magnet at high temperature, pressure, and velocity, which converts the gases' energy directly into electricity. The exhaust heat is then used to power conventional turbines. In a coal-fired MHD system—heat for the system may come from fossil or nuclear fuel—efficiency could approach 50 percent, which compares favorably with a traditional coal-fired plant with an efficiency of around 40 percent.

Scientists at Argonne Laboratories have already built superconducting magnets for MHD experiments, and one, weighing 180 tons, is among the largest superconducting magnets ever built. A 40-ton version was also built by Argonne for a cooperative research program with the Soviet Union.

The magnets do require expensive helium refrigeration systems to cool the niobium-titanium coils to superconducting temperatures. Thus, a liquid nitrogen system should be less expensive, simpler to operate, and more reliable. But some scientists are dubious. Said John Hulm:

> MHD is currently in the doghouse in the U.S. It falls in the same category as fusion. We can build superconducting magnets for them, but even with the high fields, these MHD magnets just don't work very well. At first, this technology was offered as an answer to efficiency, 50 to 60 percent versus 40 percent for coal. Now, if you want efficiency, the easiest way to get it is to build a high-temperature gas turbine, in which waste heat from it goes into a steam turbine. We can get 60 percent with that. It works because we have a tremendous basis of technology that comes from the aviation industry.

ENERGY STORAGE

Imagine the world's largest storage battery. Set in an underground trench, it is doughnut shaped, like the tokamak reactor, more than one-half mile in diameter and 20 yards thick, its supercooled superconducting wire coils bathed in 10 million liters of liquid helium. Circling endlessly at the speed of light are up to 5 million kilowatt-hours of electricity; generated at night during times of low demand, it can be stored indefinitely and siphoned off when needed during the peak daytime hours.

With such a system, utilities could retire the older, fossil-fueled power plants now used to meet demand during peak daytime hours. Nowadays, utilities often run their generators at top speed at night, using the excess electricity to pump water uphill into reservoirs. During the day, the water is released, and used to run turbines, a process that recovers only about 75 percent of the stored energy. A magnetic storage system would give the nation the equivalent of 15 to 20 percent more generating capacity because the utilities would be able to use their existing facilities far more efficiently in returning the energy stored at night. Replacing the superconductors in the storage coils with the new nitrogen-cooled ceramics would reduce the cost another 5 to 8 percent.

Unfortunately, the gigantic battery doesn't exist yet. Though superconducting magnetic energy storage was conceived in the early 1970s by professors Roger W. Boom and Harold Peterson of the University of Wisconsin, it never really caught fire. For years, the utilities doled out enough money to allow Boom to keep pursuing his research, but not the many millions necessary for a full-scale test model.

The engineering problems were formidable. The immense coil had to be braced against the enormous expansion forces caused by the magnetic field, since even the slightest movement in the structural components would generate enough heat friction to quench the superconductors, allowing stored energy to drain away. There was also the danger that large superconducting magnets might self-destruct under the force of the magnetic fields they create. Finally, there were human concerns, notably the effect such powerful magnetic fields might have on people living in the vicinity.

About all anyone had to show for the limited research effort was a prototype built by the Los Alamos National Laboratory for the Bonneville Power Administration in Portland, Oregon. The big test magnet, relying on the old helium-cooled superconductors, went on line in Tacoma, Washington, in 1984, and ran for months, storing enough electricity to power a house for about a week. It was a good newspaper story, but little else.

Then the Pentagon, more specifically the Defense Nuclear Agency of the Department of Defense, got into the act. Obviously, the military were not merely interested in building a bigger and better battery to power their air conditioners through the summer. The government's interest was fairly self-serving: in order to implement the Strategic Defense Initiative (SDI), popularly known as the Star Wars missile defense program, enormous amounts of electrical energy will be required.

Superconducting magnetic energy storage (SMES) could not only help solve the world's energy problems, according to Dr. Boom, but also "have an equally impressive impact on the national security by providing a reliable source of energy to fire SDI ground-based weapons requiring large amounts of power to be rapidly available as the need arises." These space-age weapons include particle beams, electromagnetic cannons, and antimissile lasers—all of which demand the short bursts of massive energy that a SMES system could provide. Even if the utilities built such systems for civilian consumer use, they could be tapped by the military in the event of a missile attack, and the stored energy could be easily diverted to the SDI arsenal.

In late 1987, the Defense Nuclear Agency awarded $14-million, two-year contracts to two industrial engineering teams—Ebasco Services, Inc., of New York, and Bechtel National, Inc., of San Francisco—to design and develop a SMES test model, the first step in constructing a prototype capable of storing 15,000 to 30,000 kilowatt-hours of elec-

SUPERCONDUCTING MAGNETIC ENERGY STORAGE

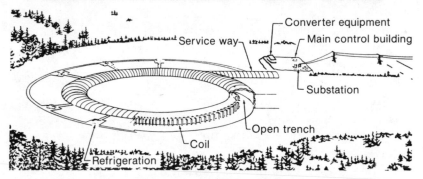

Plan for a 5,000-megawatt Superconducting Magnetic Energy Storage System places a superconducting doughnut-shaped coil underground. Energy stored in the coil would fill peak daytime demands, a task now performed by gas turbines. (*Courtesy Robert Lloyd, Bechtel Corp.*)

tricity. If all goes on schedule, the experimental system will be ready in 1992.

The superconducting storage coil, made of niobium-titanium, will be about the size of a football stadium, set in a pit cut into rock to restrain the outward pressure flowing from the magnetic field. The coil and its liquid helium coolant will be jacketed in an insulation system, the whole rig further insulated by other jackets of cold, liquefied gases such as neon and nitrogen. Stored DC current will be converted to AC and sent out onto an electric power grid. According to the experts, switching between the charging and discharging modes can be accomplished in less than a second, which means that the system could effectively limit disturbances in the electrical system that cause large-scale power outages.

TRANSMISSION

At the Brookhaven National Laboratory, joined by 430-foot superconducting cables of niobium-tin, stand two shedlike structures with twisted transformer coils jutting up over them. During eight operating runs between 1982 and 1986, those cables carried up to 660 mega-volt-amperes each, enough to power a city of a million people. It was the first demonstration of superconducting transmission lines of the future, which will carry electricity over long distances without loss from resistance. "It was only a few hundred feet long," said the DOE's Donna Fitzpatrick, "but we operated it and proved it could be done. We learned what it would cost and how to design the system."

Liquid helium cooling makes superconducting transmission lines impractical at the present time because refrigeration drains power, and Brookhaven scientists calculated that a niobium-tin line would require a refrigeration station every 10 miles. Also, most electricity travels along overhead wires, which benefit from free cooling and insulation provided by air. Superconducting cable would have to go underground to accommodate the cooling system along its entire length, which means miles of expensive trenches and insulation.

Nonetheless, the most likely large-scale application of the new ceramics is in power transmission, especially in areas where power demands are high and real estate is expensive. Copper conductors may be good, but according to the American Public Power Association, the national loss from transmitting electricity through conventional wires equals the output of fifty power plants. Copper, in fact, is so inefficient that as much as 15 percent of the energy generated must be used merely to overcome the metal's resistance. Large utilities, like Pacific Gas and Electric Company, waste as much as $200 million worth of electricity this way each year.

Substituting ceramic wire—actually, one promising design is not exactly wire as we know it, but a coating of the ceramic on a supporting tube or rod made of copper—and low-cost liquid nitrogen cooling could simplify the technology, and they would improve the economics by capturing the energy lost over conventional lines. The power that would be saved, say proponents, would also reduce the need to build more power plants—and building new plants has long been the cheaper alternative to stringing long lines. Any new plants that would be constructed could be sited in once-remote areas where there are untapped power

sources, and be part of a vast, continent-size power grid—
something not possible today because energy losses over
copper cables are on the order of 1 percent per 100 miles.
Some visionaries even imagine thousands of miles of undersea cables linking continents in one giant power network.

"Superconducting transmission lines would effectively
increase electrical generating capacity by about 5 percent,
the amount of power now lost through resistance in electrical lines," said Richard W. Weeks, director of Argonne
Laboratory's division of materials and components technology, "and would allow electricity to be carried long distances economically. We could make greater use of site-
specific sources, such as geothermal, hydroelectric, and
solar energy, located far from population centers. Nuclear
and coal-fired power plants could be built away from
populated areas."

There seems to be general agreement that for a national
distribution grid, superconducting underground transmission would be far more costly than copper wires strung
overhead. That could change, of course, if room-temperature superconductors are ever developed for practical use.
On a local scale, however, underground superconducting
wires could cut AC transmission costs by as much as 40
percent, according to a recent national survey conducted by
Argonne on the possible economic impact of high-temperature superconductors. Savings would come mainly from
reducing energy losses by about 75 percent.

Alan Wolsky, Argonne's senior energy systems scientist,
is one who believes that high-temperature superconductors
have the potential to spawn new technologies and cut
capital and operating costs in numerous applications, including the development of underground transmission
lines. He is, however, not given to grandiose predictions of
future applications, and when he discusses the new superconductors' possible impact, he invariably points out that
application will come only when present problems are
solved, the most important of which is boosting the ceramics' current-carrying capacity. Wolsky added:

In the last analysis, you're only going to see superconducting underground transmission lines if there's a reason to go underground. Like citizens' groups who don't want the landscape marred, or someone's concern about the biological effects of a magnetic field, or New York City, so densely populated that there's no room for overhead lines, and you have to go into conduits underground. But if you have to take power from the Zion nuclear plant near Chicago, you have to go above ground, it's cheaper that way, and superconductivity won't change that—until it's at room temperature. Efficiency is not the answer for underground superconducting transmission lines. Such lines might lose less than the 1 percent of electricity per 100 miles lost over conventional aerial lines, but they won't make them obsolete. But aerial lines do have problems with aesthetics—and potential health risks that may make underground lines a viable alternative. If this occurs, high-T_c superconducting lines might save 40 percent over conventional underground lines, which experience energy losses of nearly 4 percent.

13

MAGNETIC FIELDS
AND THE BODY

As THE ENGLISH satirist Samuel Butler once said: "A right way of looking at things will see through almost anything."

X-ray, or roentgenography as it is known after its discoverer, the German physicist Wilhelm Konrad Roentgen, gave doctors a right way to see into the human body, and revolutionized the diagnosis and treatment of disease and deformity. Since its first clinical application in 1895 by Canadians John Cox and Robert Kirkpatrick, who used it to detect a bullet in the leg of a patient, x-ray's penetrating gaze has laid bare the interior of the human body without cutting it open.

But as fantastic as its power is, x-ray has many shortcomings. The x-ray photographs are often too misty and generally require a highly trained eye to detect something that the layman staring at the same negative would miss. While bones are clearly visible, the soft tissue in our organs, the cartilage in our noses, and the gelatinous disks that separate our vertebrae appear only as vague shapes, if they appear at all. The same applies to our nerves, our veins, plaque in our hearts and brains, and many tumors. X-ray, moreover, can be hazardous to our health, particularly when we are overexposed to its cell-altering beams.

Fortunately, there are now better ways to assess the inner state of our bodies: "cameras" that make x-ray seem almost like a relic of the days of saddlebag doctors. These new devices rely on the dramatic advances in solid-state electronics and computer technology and on vastly improved knowledge of sound waves, infrared sources, the protons in the nuclei of hydrogen atoms, radio waves, magnetism, and in many applications, superconductivity.

There is CT, for computerized tomography, a wedding of x-ray and the computer that scans a portion of, say, the brain by revolving an x-ray tube around the head, converting the pictures into a digital computer code to make high-resolution, three-dimensional video images accurate to .9 millimeter. Differences in tissue density, between abnormal and normal tissue, are revealed; fine details of bone structure appear clearly; the location of tumors is pinpointed.

Sonography, the diagnostician's version of the very sonar that navigators use to determine the depth of the water under a ship's keel, sends short bursts of sound waves into the abdomen, then recaptures them as they bounce back, and sends them to a computer that translates them into an image—of a fetus, a tumor, a gallstone. A more sophisticated version, digital color Doppler, takes advantage of the Doppler effect—a change in the frequency of sound, light, or radio waves produced as a moving object draws near or away from an observer—to monitor the sound of blood flowing through the heart, veins, and arteries and convert it into pictures. If blood flow is normal, it has a gentle sound, and the sound picture is of a smooth current; if the blood must squeeze through an artery narrowed by plaque or some congenital deformity, the sound may be as scratchy as a worn record, and the echo image will depict an irregular flow.

PET (positron emission tomography) uses trace amounts of radioisotopes to measure blood flow through tissue and to determine if a patient's biochemical processes are functioning properly.

There are others, but three that depend on magnetism are of interest in our discussion of superconductivity: MRI, for magnetic resonance imaging; MRS, magnetic resonance spectroscopy; and MEG, magnetoencephalography. MRI and MRS use magnetic fields and radio waves to scan what lies inside the human body; MEG attempts to study activities in the brain by monitoring the magnetic fields the organ itself generates.

Nuclear magnetic resonance, or NMR, the driving force behind MRI and MRS, was discovered in the mid-1940s. A layperson hearing the term for the first time might well associate it with some new piece of audio equipment for the home entertainment center. Resonance, after all, does have a certain ring to it since it does have something to do with acoustical vibration, but NMR is not of the sort of resonance one generally thinks about.

This kind of resonance refers to the behavior of the protons—positively charged particles—in the nuclei of atoms. When a magnetic field is applied, the nuclei, which ordinarily whirl about and face in every conceivable direction, line up facing the field's poles like marchers responding to a drill sergeant's commands. In this formation, the protons still gyrate, their rate of spin, the so-called frequency, increasing with the strength of the magnetic field. But these are not just the dervishes of the atomic world, content to wobble in place in some sort of devotional exercise for the good of pure physics. The protons are capable of much more. When they are hit by radio waves carefully pulsed to match their frequency—these waves are similar to those in television and FM radio signals—the protons are nudged out of line. But only for nths of a second. When the radio waves are turned off, the protons bounce back into line, only this time, excited by the radio frequency radiation they have received, they reemit radio signals of their own, or as the physicists like to refer to it, a resonant frequency. It is somewhat like what happens when a water glass is tapped with a fork: the tapping puts energy into the glass, and some of it is returned in the vibration.

It didn't take investigators long to find out that each of these resonant frequencies was specific to one type of atom, and that the frequencies shifted predictably in the presence of other atoms, as in a chemical compound. Since each of the nuclei had, in effect, a chemical fingerprint (its spin state in a strong magnetic field), this meant that even the most minute amounts of specific nuclei could be identified in large amounts of other material. All a researcher had to do to analyze the composition of complex mixtures was to tune in on a range of frequencies and in no time he could identify all the different elements in a sample. Virtually all of the common elements—including hydrogen, oxygen, nitrogen, carbon, sodium, phosphorus, aluminum, silicon, and chlorine—can be examined by magnetic resonance spectroscopy, and it has been used for many years in biomolecular research. (It should be noted that MRS is but one of several specialized kinds of spectroscopy used by scientific investigators with different interests, from astronomers trying to determine how much the universe has expanded, to physicians who need to know the concentration of different chemical compounds in tissue.)

For instance, MRS's giant magnets are able to produce graphs showing the kinds and amounts of various biological chemicals present in particular areas of the body, and thus can identify sequences of events that may be taking place. Doctors are now using the technique to assess damage in, say, an infant's brain cells when its oxygen supply has been decreased during birth. The spectrometer's strong magnetic influence is used to measure adenosine triphosphate, a compound produced through both aerobic and anaerobic metabolism to supply cell energy. If a baby develops a high degree of acidity in its brain cells, the test will reveal the change and the need for more oxygen.

The spectrometer is also used to identify a patient's so-called anaerobic threshold, the point at which the oxygen supply to cells can no longer keep up with demands. A runner, for example, has a certain speed at which he or she can maintain a steady pace for long periods. Any significant pickup in momentum will usually lead to rapid fatigue through acid buildup in the body's cells. The threshold separates two kinds of energy available: the more efficient mechanism, aerobic metabolism, uses oxygen; the other, called anaerobic metabolism, uses another substance. Before MRS spectroscopy came along, there was no ideal way to detect the threshold, and most techniques required drawing blood samples or a biopsy, a procedure in which a tiny bit of tissue is removed for laboratory examination.

Eventually, someone came up with the idea that magnetic resonating might be applied to the human body, not only to analyze its biochemical reactions, but to image its organs and other soft tissues as they had never been seen before. The body is, after all, a conglomeration of ordinary chemical elements composed of atoms whose nuclei are capable of sending out radio signals. Also, the most abundant nuclei in the human body belong to hydrogen atoms, which happen to be the most sensitive to detection because of their outstanding magnetic makeup.

Enter MRI, which used to be called NMR, for nuclear magnetic resonance, until someone decided that "nuclear" had a rather ominous tone that could unduly frighten a patient. Unlike MRS, which gleans biochemical information from bodily tissues, MRI reads the signals returning from hydrogen atoms and feeds them into a computer, which processes them and forms a detailed, high-contrast image of soft tissues, as in the brain and spinal cord. Using the color-enhanced pictures, a physician can differentiate between normal and abnormal regions. Multiple sclerosis plaques in the brain, not visible by any other means, are seen by MRI, as are brain lesions caused by vitamin B_1 deficiency (common among alcoholics), tiny tumors hidden in the heart and pituitary gland, and torn knee cartilage. The whole process is noninvasive and painless, and in many of its applications far superior to x-ray.

But MRI equipment is expensive—around $1.5 to $2 million, sometimes much more. And it is big and bulky: 6 feet by 8 feet by 10 feet, it weighs 35 tons. The unit, which can accommodate an entire human body, includes a huge, 1.5-tesla superconducting electromagnet that generates a field thirty thousand times stronger than the earth's, liquid helium, insulation, a radio-frequency generator, and a computer. There are some twelve hundred of these giant machines throughout the world, close to a thousand of them in the United States alone. But because of their high cost, few hospitals can afford them.

The superconducting component of an MRI system is, by far, the costliest part of the unit. It alone can set a hospital back at least $500,000. The machines, moreover, use around 80 percent of all of the superconducting material now sold—miles of niobium-titanium wire are wound into each magnet—making MRI one of today's few applications of superconductivity.

If the new ceramic superconducting materials are ever fabricated into flexible wire, MRI machines and spectrographic tools could be brought into widespread use because liquid-nitrogen cooling would lower construction and operating costs, save patients up to 10 percent over current MRI charges, and reduce the complexity of operating the equipment. Machines could be made more portable, allowing them to be truck-mounted and driven to hospitals and clinics that cannot afford to buy one. But the most valuable economic benefit might come from an entirely different direction: high-T_c superconductors could increase the usable diameter of the magnet, the cocoon in which the patient fits, by 6 inches. The space is now taken up with cooling equipment and insulation, and enlarging the diameter would allow the machine to be used on the 5 percent of the population who are too overweight to fit inside, thus increasing the economic gain by lengthening the productive life of treated patients. "If people with the most problems, the obese, can't get into the hole, then what good is the machine?" asked Alan Wolsky of Argonne. "It would really be an advance if we could screen those people."

MEG, magnetoencephalography, now in a testing phase, is another potential use of superconductivity to screen the human body. Unlike MRI, which surrounds a patient with huge superconducting magnets and employs the field to excite protons, MEG relies on the tiny, ultrasensitive su-

perconducting quantum interference devices, called
SQUIDs, to detect and measure the even tinier magnetic
fields generated by the nerve impulses of the human brain
itself. These weak fields, only about a billionth of the
earth's magnetic field, cannot be detected by any other
means. But with the MEG, clinicians can learn precisely
what goes on in the brain, and where.

The idea that magnetic fields are associated with biological systems was not always an acceptable one. It had a
rather rocky start with an Austrian physician, Franz An-
ton Mesmer, who treated hysterical patients in Paris in the
late 1700s by what he suggested was an ubiquitous magnetic power concentrated in himself but what is now
known to be hypnosis, or as it is sometimes called, mesmer-
ism. Mesmer, who referred to his curative powers as animal magnetism, eventually fell into disrepute and spent the
last years of his life in obscurity.

Today, biomagnetism is a legitimate scientific discipline,
due largely to the pioneering efforts of G. M. Baule and R.
McFee, two Syracuse University electrical engineers who,
in 1963, had a colleague lie down in an open field away
from electrical interference, placed a pair of wire coils
around his body, and successfully measured the magnetic
field from his heart (the organ that produces the strongest
field, because its electrical currents are stronger). The
heart's electrical currents, which are carried to the body's
surface, originate in a region of muscle, the pacemaker,
where the energy that triggers a heartbeat is generated.
These currents provide a graphic record of heart action and
can be measured directly with the familiar electrocardio-
graph.

The eye also produces a fairly strong magnetic field—
shades of Mesmer—because of a standing potential difference across the retina, and so too does the liver, because of
iron concentration; the arm, because of a steady ionic flow
along its length; and dozens of other organs and tissues.
Each organ produces a characteristic field pattern, and
important diagnostic information might be obtained from

these fields once the technique of biomagnetic measurement is perfected.

It has been suggested, for example, that a skin cancer upsets chemical balances, which upsets electrical current, which ultimately upsets the normal magnetic field. Sensors might pick up that magnetic deviation and thus provide an accurate diagnosis without biopsy and laboratory analysis. In the case of diseases that lead to an accumulation of iron in the liver—children with thalassemia, for instance, need repeated blood transfusions, which leads to iron buildup and liver damage—a biomagnetic sensor could conceivably provide a better measure of the iron concentration than a painful biopsy. This would allow doctors to better assess whether chemical therapies aimed at binding the excess iron and removing it from the body were working.

But monitoring magnetic fields from organs like the liver is not as easy as it sounds, mostly because other nearby organs also produce their own fields and thus become magnetic contaminants. Magnetic interference may also come from an outside source. If, for example, you open a can of beans and eat them, your stomach becomes sufficiently magnetic from the fine metal particles dropped into the beans from the cut lid. Sensors could easily pick up such exogenous magnetism in the stomach, which would impede clinicians' efforts to detect, say, an early stomach cancer by measuring the stomach's own natural magnetic fields.

Such obstacles have not deterred the small group of scientists who are tracing the body's magnetic fields, primarily those that arise from electrical activity in the brain, the master computer that is, in reality, an electrical instrument linked by nerves to every part of our bodies. It is at work whenever we are afraid, hungry, thirsty, angry; when we see, smell, feel, hear, move, and speak. The messages that direct all of these actions and reactions are pulsed through our bodies over a network of billions of nerve cells, neurons, by electrical charges created by biochemicals. Put your hand on a hot stove, goes the old example, and pain nerves are stimulated, transmitting an electrical signal to the brain, which says, "It hurts." Instantly, the brain sends a message back to the hand, instructing it to pull away from the stove.

Every person has his or her own distinctive brain wave pattern, the result of electrical discharge, and it may be traced by electroencephalography. Tiny wires, electrodes, are attached to the scalp and connected to an amplifier that magnifies the electrical impulses a million times—the potential charges that reach the scalp are on the order of 5 to 200 microvolts—and translates them into squiggles on a moving strip of paper. This electroencephalogram, the EEG, can identify brain disorders and metabolic disturbances because such anomalies cause abnormal wave forms.

Every electric current produces a magnetic field, and this is so whether it occurs in your everyday, wire-wrapped electromagnet or in the brain. In the head, these fields, known as neuromagnetic fields, are found in the space surrounding the brain under the skull and are not distorted—as electrical currents are distorted—when they reach the scalp and emerge.

The first report of a neuromagnetic field was made in 1968 by David Cohen of the Francis Bitter Magnet Laboratory. A few years later, with the advent of the extremely sensitive SQUID, Cohen was able to refine his work. With ultrasensitive detectors strapped around their heads, his subjects bedded down in a many-sided metallic room designed to shield them from outside magnetic fields. The brain signals that were monitored in this way enabled Cohen to map the so-called magnetic alpha rhythm field, one of the four brain wavelengths associated with various mental states. (Alpha waves are generated almost exclusively during periods of meditation and relaxation; beta waves during anxious moments; delta during sleep; and theta in creative periods.)

Other scientists, like physicist Samuel Williamson of New York University, one of the leading researchers in neuromagnetism, have expanded on Cohen's work. Using the SQUID-based sensors, which owe their sensitivity to superconductivity, they've been able to explore and measure brain activity more fully. "MRI as applied to imaging of the human body," Williamson explained, "is an anatomic [structural] image, and what we are developing is a functional image, a functional image of the brain. A number of studies support the theory that magnetic field patterns are generally more sharply confined over the active portion of the brain than are electric potential patterns."

The magnetic technique, therefore, provides a direct measure of brain activity. The EEG, said Williamson, cannot localize neural activity; it has a problem with separating patterns from different simultaneously active sources. Also, the EEG is very much affected by intervening layers of tissue, especially the skull, which sets up a barrier of high resistivity. But the brain and surrounding tissues are transparent to magnetic fields, so that they emerge from the head without distortion. By measuring the pattern of the magnetic field around the head, Williamson can deduce where neural activity takes place inside, where neural firings are going on. "Through source localization," he explained, "it is possible to reveal significant aspects of brain function, either normal or abnormal, and we can demonstrate how function is distributed through the brain."

The sensors that Williamson uses in his studies are built to overcome and compensate for the main challenge in measuring neuromagnetic fields, the overwhelming background magnetic "noise" in the environment—noise, for instance, from the AC magnetic fields of motors, subways, and elevators. The magnetic interference from a subway system eleven stories below Williamson's lab, for example, produces a field a million times stronger than the fields of interest in the brains of his subjects. David Cohen solved the problem with his elaborate, magnetically shielded room, but such an enclosure is expensive.

Williamson's magnetic technique employs a cylindrical Dewar filled with liquid helium, inside of which are the SQUID, which gauges the magnetic field, and another key

MEASURING MAGNETIC FIELDS IN THE BRAIN

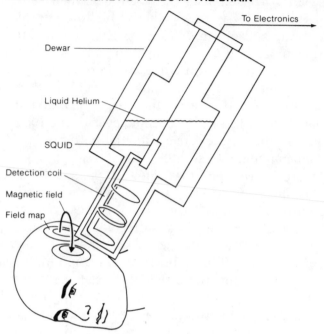

Sensor system uses superconducting coils to measure the brain's magnetic fields. (*Drawing courtesy of Samuel Williamson, New York University.*)

element of the system, the detection coil that actually picks up the brain signals. The coil, connected by leads to the SQUID, is made of three individual coils of superconducting niobium-zirconium mounted at the bottom of the Dewar, the end that is placed near the subject's head.

The field from a single neuron is far too weak to be detected outside the scalp with the present system. What Williamson is measuring is a field produced by ten thousand nerve cells working in synchronization. It sounds like an enormous number, but when one considers that 1 square millimeter of cerebral cortex—the wrinkled outer layer of gray matter, where various sensory and motor responses are coordinated and controlled—has about a hundred thousand neurons, the localizing capability of the MEG system becomes quite precise.

Williamson's scanning system allows him to aim fourteen sensors at once at the head of a subject, an improvement over earlier techniques, which used only a single sensor that had to be moved from one location to another, covering forty positions before a meaningful pattern was revealed. The process could take eight to ten hours.

Using MEG scanning, Williamson monitors various strong responses in the brain and tries to assess just how the brain sorts out, say, the sights and sounds around us before it makes a decision about what to actually see or hear. To gather such information, Williamson exposes subjects to various stimuli—pictures, sounds, a touch—while the sensors are pointed at their heads.

For example, scientists are aware that a "photograph" of tangible objects is not actually projected onto the vision-processing part of the brain. Rather, the sensory system carefully selects certain aspects of the sensory information, specific elements to be emphasized and dealt with. To find out how the brain accomplishes all of this, volunteers were shown a design of shifting bars against a solid background. By altering the speed at which the pattern shifted and monitoring the fluctuating magnetic fields picked up by the sensors, Williamson was able to locate the very cells that appeared to respond to movement, as well as the ones that actually detected patterns. Interestingly, the U.S. Air Force is examining such neuron responses in an effort to learn more about the effect of heavy workloads on pilots, the aim being to improve cockpit design.

Another study was designed to determine how neural circuits are organized to provide a perception of sound. Specifically, Williamson wanted to find whether individuals have a "tone map" across the auditory cortex—that is, whether tones of different frequencies evoke neural activity at different locations. While Williamson's subjects listened to notes of a scale, the sensors hunted for magnetic field variations, and the researchers were able to monitor the movement of nerve impulses from cell group to cell group inside the brain. It appeared that the brain assigned equal numbers of neurons to each octave of the musical scale, much like the arrangement of keys on a piano. "This ability of magnetic studies to show differences in the locations of activity," said Williamson, "is an important step toward establishing a functional map of activity across the auditory cortex."

The new high-temperature superconductors can be expected to upgrade MEG's already formidable abilities—but not in the way most people would think, according to Williamson.

> Clearly, we need a multiple-sensor system that can simultaneously measure the magnetic field at various positions over the scalp—say, one hundred positions—so that measurements could be completed in a matter of minutes. This would enable us to see patterns of activity as they shift from one area of brain to the next. The problem, though, is that we have to keep the sensors, the detector coils, very cold in liquid helium. You also need a good Dewar, and the present ones are rigid structures. It doesn't take you long to realize, when you study human heads, that they come in different sizes and shapes—round ones, flat ones, heads with corners and edges. To maximize sensitivity, you have to get those coils closer to the head, and we don't have that flexibility with current designs. If we had a room-temperature superconductor, we could do that very easily, because they'd all be on tracks that could slide in nicely and you could fit them to a child.

14

THE SUPERCONDUCTING SUPERCOLLIDER

IF THERE IS any doubt that the Superconducting Supercollider (SSC)—the gigantic, 20-trillion–electron-volt, high-energy-physics research machine that the United States hopes to have built twenty-five miles south of Dallas by 1996—will be the most ambitious technological application of superconductivity ever attempted, just glance at the formidable industrial specifications:

- 60 million feet of superconducting niobium-titanium-copper cable—requiring 600 tons of niobium-titanium alloy and 2,000 tons of copper—for the ten thousand powerful magnets that will be required to focus streams of protons
- 100,000 tons of iron and 5,000 tons of lead for the experimental detectors
- 44,000 tons of iron for the magnet yokes
- 10 million cubic feet of concrete for the underground, 53-mile-long, racetrack-shaped ring around which the protons will speed

- 110 miles of stainless steel bore tubes
- 11,250 tons of stainless steel for the magnets
- a complex array of electronics to equip the detectors, including preamplifiers, digitizing and filtering electronics for up to 1 million channels
- the world's largest liquid-helium refrigeration plant, including eleven 5-kilowatt liquid-helium plants
- an initial inventory of liquid helium of 2.4 million liters, with an annual usage of 0.5 million
- electrical equipment including hundreds of power suppliers, more than 100 miles of distribution lines, and multiple power transformers
- the equivalent of several of today's super minicomputers to acquire and monitor data for each experiment, plus the equivalent of 100 mainframe computers for off-line analysis
- $100 million worth of construction equipment— tunneling machinery, earth movers, and trucks
- $75 million for tooling—shop equipment, dies, and presses

The total cost of this titan of magnetic force, the world's largest particle accelerator: $5.3 billion. When and if it is built—Congress, at this writing, still has not approved the project—scientists will have access to the largest, costliest research tool in the world, and the most powerful instrument for probing the structure of matter.

Like a microscope or a space telescope with far greater magnification than its predecessors, the SSC, with an energy level twenty times higher than that of the 1-million-electron-volt Tevatron at the Fermi National Accelerator in Illinois, will enable scientists to explore the basic structure of matter and the nature of elementary particles. Astrophysicists will use it to probe deeply into the beginnings of the universe, into the as yet inexplicable events that occurred in those early cosmological moments. Chemists, materials scientists, and biologists—all of whom rely on synchrotron light sources, which depend on accelerators—will benefit. In short, the research frontiers set by Newton, Galileo, and Einstein will be extended, as matter, energy, space, and time give up secrets now constrained by current, imperfect technology.

There will be some practical applications as well, leading to whole new industries, among them advanced electronics and nuclear medicine. Both disciplines have already benefited enormously from the technology of particle accelerators, starting with the invention of the cyclotron, the charged particle accelerator invented in 1931 by Ernest Lawrence, founder of the Lawrence Berkeley Laboratory in California.

In a speech at the 1987 National Symposium on the Superconducting Super Collider, in Denver, one of the collider sites then under consideration, Energy Secretary John Herrington said:

> The cyclotron started a revolution in the quest to answer fundamental scientific questions that had puzzled the world since earliest times. The discovery was a crucial step in ushering in a new era of science and technology that has resulted in unimagined and rapid advances in microcircuits, lasers, medical treatments, understanding of disease and in the harnessing of the atom. Millions of jobs and a sizable portion of our nation's gross national product—some estimate as much as a third—can be traced to developments and industries whose origins come from basic research aimed at understanding the atom. Television sets contain a particle accelerator, an electron gun whose stream of particles is modified by electric and magnetic fields and scanned across a phosphorescent screen. The electron microscope is an electron accelerator of remarkable precision. Integrated circuits, which are at the heart of all modern electronics, from computers to car radios, are manufactured by ion implantation using spe-

cially designed accelerators. Synchrotron light sources, and novel techniques for medical diagnosis, have related roots.

In light of such an impressive litany, it is no wonder that Herrington referred to the SSC as "this national treasure." (He has also called it "this crown jewel of science.")

Though some scientists feel the SSC's cost will drain scarce federal funds from other sciences, and politicians grumble that the taxpayer will not take kindly to such an expensive project, it is hard for a mere mortal not to be humbled by a machine that might eventually bring about some wonderful practical things and unlock the secrets of time and the universe.

President Reagan approved the Superconducting Supercollider as an investment in knowledge that would pay big interest later on. "If automotive technology had progressed as fast and as far as superconductor technology has in the last twenty years," he said in remarks before a group of SSC supporters, "a Rolls-Royce today would cost less than three dollars, get three million miles to the gallon, and six of them would fit on the head of a pin."

The SSC is a scaled-up version of a particle accelerator, a machine that relies on electricity and magnetic fields to boost subatomic particles, generally positively charged protons and negatively charged electrons, to velocities near the speed of light. At start-up, devices outside the machine generate these subatomic particles and release them into the accelerator's vacuum chamber, which prevents them from being scattered by air molecules. Inside the chamber an electrical field accelerates the particles—mechanical or other forces would be too weak—and powerful magnets produce magnetic fields that focus and guide these tiny aggregates of matter and keep them moving faster and faster on a circular or linear path, depending on the type of accelerator.

As they near the speed of light, the "sound barrier" of high-energy physics, the particles take on extremely high amounts of energy, measured in electron volts (eV), the

unit of energy an electric field bestows on a charged particle. By way of comparison, one particle in a sunbeam
carries 1 eV of energy, and so does one electron running
from the negative to the positive poles of a standard flash-
light battery. The Fermi Tevatron gives its protons up to 1
trillion electron volts (TeV), and a machine under construction in the Soviet Union, 2.2 trillion. The particles in the
SSC will have twenty times the energy of those in the
Tevatron.

When they are fully loaded with energy, the fast-moving
particles may be aimed at a target—the nucleus of any
atom, say—to smash it (hence the term *atom smasher*) and
change the atom of one element into that of another. The
change is called transmutation, something akin to what the
old alchemists were attempting when they tried to turn
base metals into gold.

In the SSC, which is classified as a proton-proton collider, the ten thousand superconducting magnets will guide
two beams of protons in opposite directions around the 53-
mile circuit of the tunnel. When the particles collide head
on, their kinetic energy (velocity) is converted into mass in
the form of new and exotic subatomic particles, which, as
they fly off in all directions, may exist for less time than it
takes to wink an eye. It is like the one great burst of energy
that spreads a fireworks display across the sky for a brief
moment before leaving a trail of flickering embers that
quickly die out.

Such head-on collisions result in far higher particle energies than can be achieved with straight-line linear accelerators that fire a single accelerated beam at a stationary
target. For example, according to a 1985 report on accelerator technology by the Lawrence Berkeley Laboratory, one
of the centers that has been working on high-field superconducting magnet design for the SSC, it would take a
fixed-target accelerator with a beam energy of 2,000 TeV
and a circumference of more than 8,000 miles to equal the
energy available in a head-on collision between two protons that have been accelerated to 1 TeV. Because of such
advances in accelerator technology, the energy range that

is accessible to scientists had increased a millionfold in the past fifty years, while the unit cost (the cost per electron volt of beam energy) has decreased. And it is only at such high energies that scientists can create the ultrasmall particles—atomic debris, really—for study.

When the supercollider is in operation, the particle collisions that occur are like microscopic recreations of the fiery "big bang" that marked the creation of the universe, and the data gathered by computerized detectors around the collision zone is expected to provide scientists with clues to the most fundamental structure of matter, and the nature of the forces that lie in the atom's innermost, unprobed space.

"It's a bit like finding out how cars work by smashing them together and seeing what falls out," physicist Carlo Rubbia of the European Laboratory for Particle Physics (CERN) has said. "But in particle physics, when you smash two cars together, you get 20 or 30 new cars, or even a truck or two. We're repeating one of the miracles of the universe—transforming energy into matter."

At the level of matter studied by Rubbia and the other particle physicists are the quarks, elementary particles that make up the very protons and neutrons that form atoms. But are these infinitesimal building blocks of matter truly fundamental, or are they too composed of even smaller particles interacting according to some highly ordered scheme? The higher energies and the more violent collisions produced by the SSC will, it is expected, answer those and other questions. It may resolve the one big question that all theoretical physicists are asking these days: is there a Theory of Everything, one that sees the four forces of nature—electromagnetism, gravity, and the so-called weak and strong nuclear forces—as just elements of one fundamental force, the same unified force that existed during the big bang, when space, time, and energy were created, and which separated when the universe cooled down?

In an attempt to prove their ideas, physicists build bigger and more powerful colliders to hunt for basic particles like the Higgs particle, and the "top quark," a shadowy entity that, like the Abominable Snowman of the Himalayas, leaves tantalizing tracks and an occasional fleeting silhouette before disappearing under the category of a false sighting. With a machine as powerful as the SSC, the top quark, which is believed to lurk in an eV range beyond that of today's accelerators, may one day, after some spectacular proton collision, emerge from the nuclear rubble to reveal the one, ultimate law that governs the creation of all nature.

Such are the goals that justify the construction of a machine as expensive and as complex as the SSC. But a key question continues to nag the people who would build the supercollider and the physicists who would use it: would it not be better to wait until the high-temperature superconductivity materials are perfected and use them instead of the older liquid-helium-cooled windings for the magnets?

In the SSC and in other accelerators, the electromagnets are used to bend the streams of particles and keep them on the rails, so to speak, until they near the speed of light. The magnetic fields are also used to store particle beams and to detect and identify elementary particles. Two kinds of magnets are generally fitted into the system: dipoles to contain and bend the beam, and quadrupoles to focus it to a narrow diameter.

In existing machines, the powerful magnetic fields produced in an accelerator's electromagnets come from high-voltage electric current circulating in copper coils wound around an iron core. But electrical resistance in the wire limits the flow of current and also produces a huge amount of heat that constantly has to be bled from the system. Because of those limitations, conventional electromagnets cannot create the field strength required for a machine as powerful as the SSC.

The SSC, on the other hand, would use magnet coils wound out of superconducting wire, in which an electric current can circulate endlessly, without encountering any resistance. Moreover, the superconducting magnets can generate immensely powerful fields because current could be made to flow through the wire windings without the high voltages needed to push current through conventional windings. Also, the bubble chamber—the chamber in which the path of an ionizing particle is made visible by a string of vapor bubbles—at CERN, for instance, requires a power of 70 megawatts with conventional coils; with superconductors, the device would consume less than a megawatt.

But even the magic of superconductors has its limits. Magnet designs involve relatively sharp bends in the conductor, a requirement which has led to difficulties with such materials as niobium-tin, a great superconductor but brittle compared to niobium-titanium. Cooling the conventional superconductors to extremely low temperatures is also a problem. And there is the propensity of the superconductors to lose their superconductivity—to go normal— in the presence of strong magnetic fields.

Niobium-titanium in the magnet windings is one solution to some of the difficulties, except for the necessity of cooling it with liquid helium. Such windings are, or should be, able to sustain a field of up to 12 tesla or more. (To recap, a tesla is equal to 10,000 gauss; the earth's natural magnetic field is around 0.5 gauss; and a magnet that pins notes to a refrigerator door has a field of only a few hundred gauss.)

Would the new superconductors, the ceramics, make even more of a difference? It seems logical to assume that they would. First of all, the accelerator magnets would no longer have to be cooled with costly helium to 4.2° K but would operate at liquid nitrogen temperatures, or even at room temperature. The windings would generate much stronger magnetic fields with almost no loss of current, and at far less cost than the conventional superconductors. Also, redesigning the SSC to take advantage of the more powerful ceramic-wound magnets could mean a tenfold reduction in the number of magnets and, thus, the length of the accelerator ring. With such a scaled-down machine, the land area required to contain it would be only about 1 percent of what is now planned—again greatly reducing the cost. Nothing would be lost, experimentally speaking, from a size reduction, say the scientists who favor waiting, and the money saved could then be channeled into other scientific research.

All of that is true. But despite the limitations of the superconducting materials available now, it would seem to be imprudent to delay a project as important as the SSC until higher-T_c materials become practicable. It would be like putting off marriage until the ideal mate comes along. He or she may show up, but who can tell when? According to a 1987 report in the journal *Physics Today*, "It took nearly 20 years from the discovery of superconductivity in niobium-titanium to develop stable magnets using this material." Moreover, redesigning the SSC would mean higher-field magnets, which in turn would require stronger supports to withstand the enormous electromagnetic forces; the new design could take years, and that, coupled with the stronger support structures, would add considerably to the already high cost. Putting the SSC off, then, would be disastrous, if one believes in the project. If the superaccelerator is to be built, it should be built around the conventional superconductors, with upgrading in ten or fifteen years. By then, the way the ceramics work will be better understood, and their value proven beyond any doubt.

The real question may be whether to build the SSC at all, and on that there is still considerable disagreement among scientists. Some agree that the project should be killed outright because it would take money from small science and hand it to big science. Others are opposed because they believe that high-energy fundamental particle physics, in the words of Princeton's Philip Anderson, "Has become so fundamental as to be almost irrelevant, even to the rest of science." Anderson added that if lack of the right accelerator were really going to kill high-energy physics, "I must say it is better off dead." (Anderson's opponents point out that he opposed the last generation of accelerators, which led to the discovery of quarks, among other subatomic particles.)

Even physicists who fully support the research that may be done with the SSC question whether the SSC as presently designed is the best machine for the job. "A year ago, the answer would have been an unqualified yes," said one letter writer in *Physics Today*. "Today, the answer must be 'maybe not.' Why? Because of advances in superconductor technology."

Still others warn that the argument that society will inevitably benefit from physics research—in the form of new products and entirely new industries—does not always hold water. One who has taken that stand is John F. Waymouth, of GTE Electrical Products in Danvers, Massachusetts, who recently wrote:

> As I reflect back on what physics research has provided to society in the past, I am struck by the fact that not all physics research is uniformly productive of economic benefits. In my own mind, I have divided physics into three basic areas: electron-volt (eV) physics, in which energy exchanges on an atomic, molecular or electronic scale are less than 100,000 volts; MeV-GeV physics [MeV, for mega electron volts, refers to millions of electron volts; GeV, for giga electron volts, to the billion range], which primarily involves nuclear and subnuclear particles; and high-energy physics, covering GeV to TeV and up, involving the structure of subnuclear matter.
>
> Out of eV physics have come electricity and magnetism, telegraphy, telephony, the electric light and power industry, stationary and propulsion electric motors, radio, television, lasers, radar and microwave ovens, to name just a few. In short, it is the core science of the modern world.
>
> X-rays and the resulting medical physics industry were the high-energy physics of their day, but fall within my definition of eV physics. Digital computers arose from the computational

needs of MeV physics, but the technology of satisfying those needs came entirely out of eV physics; microminiaturization of those computers for space exploration was also accomplished by eV physics, resulting in the capability to put computing power undreamed of by John von Neumann in the hands of an elementary school child. [Von Neumann was the brilliant Hungarian-born U.S. mathematician who formulated game theory, the application of mathematical logic to decision making in games, politics, commerce, and warfare, and who devised the high-speed computers that played an essential role in the U.S. development of the hydrogen bomb.]

Moreover, eV physics has been the core science in the training of generations of engineers who have invented, developed and improved products in all of the above areas. It is, in addition, the core science in the extremely exciting development of understanding the detailed processes involved in chemical reactions, and the ultimate understanding of biological reactions and the life process itself. Every single member of our society has been touched in very substantial ways by the accomplishments of eV physics, and many of them are fully aware of it.

MeV-GeV physics has given us radioisotope analysis, a substantial portion of medical physics, and nuclear energy (which a significant, vocal minority of our society regards as an unmitigated curse instead of a blessing). High-energy physics has to date given us nothing.

Undeterred by such powerful arguments and buoyed by the power of superconductivity, the SSC proponents press on. Under Secretary of Energy Joseph F. Salgado has argued that the SSC is essential if America is to maintain leadership in high-energy physics, leadership that has been based on successive commitments to more powerful and sophisticated accelerators, into the twenty-first century. "The current U.S. frontier accelerators will have explored much of the science accessible to them by the mid-1990s," he told the House Science, Space and Technology Committee in 1988, "and the SSC is technically achievable by the mid 1990s. . . . A delay could represent a tremendous lost opportunity for the Nation's competitiveness and an abdication of our half-century of leadership in high-energy physics research. By the middle of the 1990s, current accelerators in the United States will almost certainly be eclipsed by the newer European and Soviet world-class particle physics machines scheduled to go on-line at that time."

There, for the time being, the Superconducting Supercollider stands, in concept if not yet in its mammoth underground tunnel.

15

JAPAN VS. AMERICA—THE RACE FOR THE FUTURE

DEFENDING THE GOVERNMENT'S Superconducting Super-collider program back in 1987, Energy Secretary John Herrington argued that it would be an invaluable research facility for over a hundred American university-based research teams of bright young scientists and engineers bent on unlocking the secrets of nature. "The SSC will," he said, "ensure United States leadership in high-energy physics research into the twenty-first century."

But is that enough, given the less than enthusiastic attitude of some scientists toward high-energy particle physics? Even more important is the increasingly prevalent belief that the SSC will provide no payoffs in the practical applications of superconductivity. True, accelerator science has led us to rapid advances in microcircuitry, new medical treatments and diagnostics, and countless other technological benefits, but America's track record in accelerator research is not an especially convincing argument, nor is it a valid predictor of the potential applications of superconductivity. The SSC might need superconducting magnets so it can probe inner atomic space—and that in itself is a valuable application of the electrical phenomenon. Nonetheless, because the gigantic atom smasher is really a scaled-up version of an ordinary accelerator, it is highly unlikely that it will lead to practical benefits as valuable as integrated circuits, lasers, and the electron microscope, all of which are fruits of conventional accelerators.

In fact, superconductivity's more important role has little to do with the SSC and the hunt for an elusive top quark. We have superconductivity in hand, have had since 1911, and other countries, notably Japan, are poised to commercialize superconductivity technology.

Though the new materials are not yet ready for commercial exploitation, they soon will be appearing in a range of new products, given the pace of investment in product research that goes on in Japan. Moreover, if the Japanese success with magnetically levitated trains is any indication, even the "old" superconductors can be made to perform new tricks. Superconductivity already works, and works very well. Shipboard engines and commercial electric power generators can be driven with windings of old-fashioned niobium-titanium. NMR is no dream, but a reality. Fusion, which will come, and giant energy-storage batteries can employ the same conventional superconducting wire to generate the magnetic fields required to make them useful.

But things can be better, and the taming of high-temperature superconductivity will assure that. We have seen how the new ceramics would affect the power, transportation, medical, defense, and electronics industries. The commercial applications will without question have a revolutionary impact on society. And, perhaps, the most important impacts have not yet even been anticipated.

But to make it all happen requires a commitment—and despite the U.S. lead and expertise in basic research, American companies may have already fallen behind foreign competitors in the race to move high-temperature superconductivity into the marketplace. Although IBM, AT&T, Westinghouse, and a few others have made enormous strides, the prospect is bleak.

According to a comprehensive report from the Congressional Office of Technology Assessment (OTA) in 1988, American managers, under pressure to show short-term profits, have generally preferred to wait and see, looking for safe bets. Instead of aggressively examining possible applications of superconductivity, said OTA, U.S. firms plan to take advantage of developments as they emerge from someone else's laboratories, or buy into emerging markets when the time is right. Unfortunately, the agency added, such reactive strategies have seldom worked in industries like electronics over the past ten to fifteen years, and many American companies seem to have forgotten how to adapt technologies originating elsewhere.

"The education of American engineers," the OTA report said, "seems increasingly divorced from the realities of the marketplace and the factory floor." The education of U.S engineers is further hampered by the fact that high school students have become less and less interested in engineering, and in science. While American students comprise some 90 percent of all undergraduate engineering students in our universities, they make up less than a third of all postdoctoral students—the rest are foreign-born. The Japanese are well represented on U.S. campuses: according to the Japan Center at the University of North Carolina, some fifteen thousand Japanese students and researchers show up at our institutions every year.

It is not a new story. U.S. competitiveness in both smokestack and high-technology industries has been slipping for years. True, federal R&D dollars help create a vast pool of technical knowledge that the private sector, including foreign firms, can draw on. But beyond that, U.S. technology policies have relied heavily on indirect incentives for innovation and commercialization by industry. "This approach, leaving R&D largely to the mission agencies, trusting to indirect policies to stimulate commercialization, worked well in the earlier postwar period, when American corporations were unchallenged internationally," said OTA's report. "On the evidence of steadily declining competitive ability across much of the U.S. economy, it no longer works well enough. In recent years, many U.S. companies have had trouble turning existing technical knowledge into successful products and processes, and getting new technology out of the laboratory and into the marketplace."

On paper, the United States seems to be riding high on superconductivity. Taken together, federal agencies spent some $160 million for superconductivity research and development in 1988, over half of that on the new materials, the rest on low-temperature superconductivity. Moreover, the U.S. contribution to high-temperature superconductivity research and development alone is substantially more than the Japanese government has budgeted for high-temperature superconductivity and low-temperature superconductivity together. And yet, the short-term view that is fostered in the United States by our financial markets could well put our companies far behind the Japanese within three years.

There are several reasons for that gloomy forecast. One of the most important is that of the $95 million budgeted for high-temperature superconductivity R&D, $46 million went to the Department of Defense, and $27 million more to the Department of Energy. The National Science Foundation, dedicated to public-interest research to improve economic growth and competitiveness, spent $14.5 million. Virtually all of that money was funneled into the university system—a system that the United States depends on far more than do other industrialized economies to conduct the basic research that precedes commercialization of new concepts like high-temperature superconductivity.

Much of the Defense Department's research allotment, however, is spent on specialized applications in defense systems, including the Star Wars shield, with limited potential for any commercial spinoff. Defense missions, according to OTA, shape even the basic research supported by the Pentagon. The Department of Energy's share is distributed mostly to the national laboratories, which have yet to demonstrate the ability to transfer technologies rapidly and effectively to the private sector.

On the other hand, research carried out in industrial labs has, almost by definition, a practical orientation. So does engineering research in universities and nonprofit labs. Nonetheless, although federal dollars could support an enormous technology base that the private sector could build upon, the U.S. government has not seen fit to provide assistance for commercialization, nor is there any policy or tradition for that kind of support. Compounding the problem is that few high-level managers in American firms have technical backgrounds, and they may not fully appreciate the role of R&D in business strategy and international competition. "To executives fighting a takeover," says OTA, "research may look like a luxury; after a merger, it may seem expendable."

It is a far different story in Japan. The government, which regularly targets specific markets, patiently supports industry through programs for strengthening research capabilities in universities and national labs, and joint R&D programs. In the United States, industry spends about as much as government on high-temperature superconductivity R&D—but in Japan, industry is outspending government.

The reason is clear: Japanese firms place high priority on technology as a competitive weapon and see superconductivity as a major new opportunity that could set the pattern of international competition for the twenty-first century. Consequently, they have made huge commitments of scientists and funds to pursue both basic research and applications research in tandem. Japanese companies employ more chemists and engineers for superconductor work (and fewer physicists) than do the Americans, and a far wider range of companies are involved—steel and glassmakers, for example, who see their industries declining and want to diversify, as well as chemical producers and electronics manufacturers. Said the OTA report:

> Japanese managers see in HTS a road to continued expansion and exporting, and are willing to take the risks that follow from such a view.

For years the claim was common that Japanese firms got a free ride from U.S. R&D. More recently, Americans have realized that Japanese corporations have no need to imitate or to be followers. They have highly competent and creative technical staffs, fully capable of keeping up or taking the lead in fields ranging from automobile design to gallium arsenide semiconductors, opto-electronics, and ceramics. Giving the Japanese the credit they deserve has intensified U.S. anxieties over commercialization. Only in science, in basic research, do Japan's capabilities remain in question. For the Japanese, HTS presents an opportunity to show the world, and themselves, that they can be leaders there, too. Japanese R&D managers, almost unanimously, see HTS as a revolutionary technology, one that promises radical change. Skepticism, common in the U.S., has been rare in Japan. The Japanese see applications coming relatively quickly. Japanese managers find it strange that American companies believe they can track a technology's development, waiting for the right time to begin product development, without actively and aggressively pursuing that technology in their own laboratories. [Many Japanese companies, on the other hand, did just that not so many years ago.]

But even if the United States continues to make many of the scientific discoveries, and even if American firms were to be the first to introduce innovative new products based on superconductivity, Japanese companies could still take the lead in commercialization, as they have done in consumer electronics. Relatively quickly, the competition would come down to a matter of engineering skills and manufacturing know-how—the latter being especially important since the new high-temperature materials are difficult to work with. And in these areas, the Japanese companies have tremendous strength. It is in those same areas, on the other hand, that American industry seems to have lost out.

"The Japanese production advantage is much less a product of better tooling or better technology than of better implementation," observed Michael Borrus, deputy director of the Berkeley Roundtable on the International Economy, at the University of California. "Among other things, the Japanese advantage has been obtained in the careful management and operation of its production processes, in the close working relationship between producer and supplier, in rigorous quality control and in the continuous tinkering that provides slow but sure improvement. The lesson for American companies is that the greatest gee-whiz technology in the world will not substitute for competence in these areas."

Mohammad Sadli, professor of economics at the University of Indonesia, may have offered another clue to that special Japanese expertise a few years ago when he observed: "Japanese technicians have a reputation for being very diligent, very responsible for their work. You don't have to command them. They are driven by their own sense of responsibility. They are also very alert in respect to technologies—small technology, factory-level technology. They always try to find ways and means to improve the workings of machinery or the process, saving raw materials, or saving energy, or improving quality and what not. These are not things found in textbooks. The Japanese do it because they are probably driven by an inner inclination."

As a case in point, OTA singled out the Josephson junction (JJ), a superfast, superconducting electronic switch. Because of their low level of power dissipation, these devices could be packed more densely into a computer, making it incredibly fast and powerful. As OTA told it, three U.S. firms—AT&T, IBM, and Sperry Univac (later merged with Burroughs to form Unisys)—pursued JJ research and development for computer applications, and each made significant technological contributions. Beginning in the

1960s, more than ten years of research at AT&T's Bell Labs produced a better understanding of the physics behind the devices. IBM took it a step further, building a prototype of the circuitry for a complete computer and exploring fabrication methods for JJ logic and memory chips. Sperry zeroed in on junctions made from niobium and niobium nitride (instead of the lead alloy used by IBM) and developed processing methods for high-performance, all-niobium circuits.

By the early 1980s, however, all three had scaled back or abandoned the work. AT&T quit after deciding that the technological hurdles to practicality were too high. Sperry pulled out partially but continued research on JJ sensors in its Defense Systems Division. IBM, with the most ambitious program, was spending $20 millon a year by the early 1980s, $5 million of it coming from the National Security Agency. Although the NSA wanted the work continued, IBM cut back sharply, ending research on a working computer after its Yorktown Heights lab was reorganized. Logic chips based on IBM's experiments worked well, but the memory chips did not. A new management team estimated that improving the memory would add another two years to the schedule, and by that time, according to the team's reasoning, continuing progress with more conventional silicon and/or gallium arsenide chips would make it difficult for a JJ-based machine to offer any compelling advantages in speed or processing power.

Before IBM ended its program, it came close to an agreement with Sperry for joint development of Josephson technology. IBM had the most advanced designs but was struggling to fabricate them, while Sperry had proven processing skills. The agreement took almost eighteen months to prepare and had apparently cleared the antitrust hurdle after NSA proposed taking the project under its wing. But the agreement was never consummated because Sperry management decided to decentralize its R&D among its operating divisions and reassigned its JJ computer group to the Defense Systems Division—a reassignment that key technical employees declined.

The Japanese approach to the Josephson junction was far different. R&D for the project was carried out by a consortium of large and small companies under the stimulus and guidance of MITI, the powerful government Ministry of Trade and Industry. The most visible agent and supporter of Japanese industrial policy, MITI is the advocate of national projects, guiding light of research and development, and operator of its own experimental facilities, like the Electrotechnical Laboratory (ETL), which has been deeply involved in superconductivity research since the 1960s. Contrary to what Westerners generally believe, MITI does not issue directives to industry, although it exerts a powerful influence on corporate decisions. At least a third of the members of MITI's Advisory Committee on Superconductivity Industrial Technology Development come from the private sector. More than a hundred Japanese corporations belong to the Science and Technology Agency's newly formed Shin Chodendo Zairyo Kenkyukai (New Superconductivity Materials Research Association), better known as the Superconductivity Forum, part of the Science and Technology Agency. (STA is another government group supporting high-temperature superconductivity.)

Spurred by IBM's advances, Japan's Electrotechnical Laboratory started research on the Josephson junction in the 1970s. Nippon Telephone & Telegraph also cranked up a program that is now one of the biggest in the world: forty scientists working on device designs and fabrication. In 1980, MITI established a supercomputer project with a ten-year budget of $100 million and the goal of building an enormously powerful machine by 1990. The Electrotechnical Lab's ongoing effort, said OTA, was absorbed into this new initiative and expanded, with MITI supporting JJ research at Fujitsu, NEC, and Hitachi. OTA observed:

> The Japanese clearly benefited from U.S. R&D on
> Josephson devices. Bell Laboratories scientists
> published their findings widely and exchanged
> information on a regular basis with colleagues in

the United States and Japan. During the eight
years of Sperry's active involvement in research
on JJ-based computers, the company's leading
researchers visited Japan several times—at
MITI's expense.

Sperry and IBM technology laid the groundwork for nearly all the Japanese developments in
JJ computing. Fujitsu's technology, for example,
has been based almost entirely on Sperry's original designs and fabrication methods. More recently, U.S. visitors to Japan have been impressed with the advances in JJ logic and
memory circuits emerging from Japanese laboratories, especially with the work on manufacturing techniques. American firms made a good deal
of technical progress on JJ-based computers before largely abandoning the field. The Japanese
continued, in part because of MITI's push, and
they too have advanced the technology. It is too
early to say whether the Japanese work will
eventually yield a faster and more powerful computer based on Josephson electronics. But it has
become clear that, technologically, the Japanese
have made considerable headway. Fujitsu and
NEC have demonstrated LSI [large-scale integra-
tion] chips containing thousands of JJ's. Both
firms are within reach of a simple computer
based on a 16-bit JJ microprocessor, and seem
likely to reach this goal by their 1990 target
dates.

It is, of course, too early to tell whether Japan, because of
its persistence with low-temperature superconducting JJs,
will gain an edge over the United States in developing
high-temperature superconducting JJ electronic devices
that could be used in advanced radar systems as well as in
faster computers. They may find, said OTA, that any
margin of improvement might not be enough to compete
with silicon or gallium arsenide—a discovery that would

mean a technical success and a commercial failure. Some scientists, in fact, believe that high-temperature superconducting JJ devices will never prove technologically useful.

Even if the Japanese do not gain an edge in Josephson junction application—and even if the JJ doesn't live up to its promises—there still are important lessons to be learned from the Japanese style, lessons that go far beyond JJ application, to the commercialization of high-temperature superconductivity itself. This is an endeavor in which both Japanese and U.S. firms originally started out even, and it is not enough, these days, to fall back on the old argument that the West leads Japan in science. Superconductivity remains one of Japan's foremost research strengths, and in engineering, said OTA, Japanese companies have long proved their capabilities. They may also be coming on strong in other areas. "The Japanese regard themselves as ahead of Europe in most sectors except some biological sciences," observed Dr. C. Clive Bradley, a British physicist who served as his government's counselor for science and technology at the British embassy in Tokyo. "But this could change: perception of low standing is likely to encourage new Japanese efforts."

OTA concluded:

> Corporate managers in the United States and Japan look at the world differently. In seeking strategies for profits and growth, they make different kinds of choices, set different priorities, because they operate in contrasting economic, political, and social environments. Companies that do business on a global scale—IBM, Du Pont, Nippon Steel, Hitachi—may have much in common in their view of the world, but there are important differences between them as well. It may be a cliché to say that Japanese firms put more weight on growth and market share than on short-term profits, but it is true, and it makes a difference in R&D strategies, business plans—the entire array of competitive choices. The U.S.

startups, financed with venture capital, that sprang up during 1987 have no counterparts in Japan. Nor do the small low temperature superconductivity specialists. Japan's joint government-industry R&D projects—a fixture of that country's industrial and technology policies—have no counterparts here. . . .

A few large American companies are pumping substantial resources into high temperature superconductivity. But many other U.S. firms—with the resources to pursue HTS if they wished—have taken a wait and see attitude. They may have a few people working on HTS R&D, but mostly just to keep track of the technology. Most of the effort in the U.S. is going toward research. American managers believe that HTS should remain in the laboratory until more scientific knowledge is at hand.

Perhaps a dozen large, integrated Japanese multinationals—manufacturers not only of electrical equipment and electronic systems, but of ceramics, glass, and steel—are pursuing multi-pronged R&D strategies in superconductivity. As in the semiconductor industry, these resource-rich companies could prove potent rivals for smaller American firms hoping to stake out a position.

Japanese companies are conducting research but also thinking about applications. They are putting more effort than U.S. firms into thinking what HTS might mean for the company's strategy. In general, Japanese managers believe that HTS is closer to the marketplace than do American managers. They also see HTS as a means of creating new businesses, while American managers are more likely to view it in the context of their existing businesses.

Managers in the larger American companies believe that if HTS takes off, they will be able to catch up or buy in. Japanese managers want to move down the HTS learning curve in real time. They believe that advantages established now will last. Scientists, managers, and venture capitalists involved in the HTS startups in the United States believe the same thing, but they are few, small, and weak compared with the Japanese companies.

Taken as a whole, the U.S. approach—driven by the need to show financial paybacks in the short term—could leave American industry behind Japan within a few years. Such an outcome is not assured. HTS could languish in the research laboratories. Or HTS could evolve like the laser industry—never quite matching the expectations of the enthusiasts, driven heavily by military needs, lacking the revolutionary impacts of the computer or the semiconductor chip.

On the other hand, HTS could grow and spread like the digital computer. Computers—especially the microprocessors and single-chip microcomputers found in microwave ovens and TV sets, banking machines and machine tools, Chevrolets and 767s—have penetrated innumerable products and manufacturing processes. The same could eventually happen with HTS technologies.

American companies, by and large, have taken the conservative view; Japanese companies have taken the optimistic view. If technical developments in HTS proceed as swiftly over the next two to three years as they did during 1987, then Japanese companies that have been laying the groundwork for commercialization will be in a stronger position.

Laying the groundwork requires patience, a quality that the Japanese have in abundance and that Americans sometimes lack. A prime example is the Japanese electronics industry, the world leader in consumer gadgetry and semiconductor chips. Japanese subsidies to their computer industry go back as far as 1952.

Perhaps the best object lesson for impatient American superconductor managers is a story the Japanese love to tell. Japanese swordsmiths of old, it is said, could master the technique of achieving optimum temperature and heat control only after a lengthy period of apprenticeship. Once, desiring to shortcut the process and learn the temperature his master was using to forge a sword, an apprentice stretched out his hand close to the red-hot blade. The master drew a knife from his sash and sliced off the apprentice's hand. If the United States is to harness the enormous potential of high-temperature superconductivity, it has to commit itself to a long haul. As IBM's Praveen Chaudhari once put it, warning that it would not be easy to transfer technology from the lab to the storefront, "You just can't say 'I've invented a new gizmo,' and throw it over the wall."

Still, by the end of 1988, there were some hopeful signs that the United States was getting its act together. In October, Argonne National Laboratory and American Superconductor Corporation, in Cambridge, Massachusetts, signed an exclusive licensing agreement that will enable American Superconductor to develop and market a technology, developed at Argonne, that could well speed commercial development of high-temperature superconductors. The method involves coating a wire with the right proportions of barium, yttrium, and copper, then heating the wire in oxygen to oxidize the coating. Said Argonne Director Alan Schriesheim: "It is vital to America's international competitiveness that new superconducting technologies be transferred quickly from national laboratories to industry. This agreement demonstrates that public and private sectors can work together."

Also, in November of 1988, the U.S. Patent Office indicated that Massachusetts Institute of Technology would be awarded a patent on a process for making flexible ceramic superconductors. The process combines the ceramic with a metal such as silver, gold, platinum, or palladium; the composite is fairly flexible, more resistant to oxidation, and more easily connected to electrical sources than previous materials.

"There's no doubt these materials are going to have a very large impact on society," observed MIT professor Gregory Yurek, one of the composite's developers. While its first uses would probably come in computer shielding, small motors, and satellite sensing devices, future applications could be more varied. Said Yurek: "We're talking about powerful desktop computers, medical imaging, geological exploration, power transmission, transportation. Even the popular idea of levitating trains is not unrealistic—it will in fact happen."

GLOSSARY

absolute zero: Believed to be the lowest possible temperature, the temperature at which molecular motion stops and at which heat is completely gone. $-460°$ F.

anisotropic conductivity: Describes conductivity in different directions of current flow.

annealing: Heat treatment of a material to soften it and make it easier to process.

BCS theory: The modern theory of superconductivity. Named after John Bardeen, Leon Cooper, and John R. Schrieffer.

Bitter magnet: An ironless magnet consisting of stacked layers of pierced copper disks. Named after Francis Bitter of the Massachusetts Institute of Technology.

bubble chamber magnet: A device to detect ionizing radiation.

CBA Palmer magnet: The predecessor of most superconducting accelerator magnets. Named for Robert Palmer of Brookhaven National Laboratories.

cermet: A composite material made by mixing, pressing, and sintering metal with ceramic.

chains: One-dimensional links of alternating copper and oxygen atoms that form the latticework of a ceramic crystal.

Cooper pairs: Electrons running in tandem through a superconductor, according to the BCS theory.

critical current density: A measure of how much current a superconductor can carry.

critical magnetic field: The maximum magnetic field that allows superconductivity to occur above which a material's ability to superconduct decreases.

critical temperature: The temperature at which a conductor superconducts and above which it stops superconducting. Resistance drops to zero at the critical temperature.

cyrostat: A low-temperature thermostat.

cryotron: A tiny switching device made of superconducting wires, used in computers.

Dewar jar: A vacuum bottle used to contain liquid nitrogen and other supercooled gases.

dipole: In an accelerator, a magnet that contains and bends a particle beam.

doping: The process of incorporating small amounts of certain chemicals into another substance.

dynamic compaction: Explosive welding in which materials are bonded by shock waves from a controlled detonation.

electromagnetic thrust: The principle by which a magnetic field plus an electrical current produce a linear force.

electromotive force (emf): The voltage produced by a generator or a battery.

electron volt (eV): A unit of the energy induced on a charged particle by an electrical field.

exciton: A subatomic particle. An excited state of an insulator or semiconductor which allows energy to be transported without transport of electric charge; may be thought of as an electron and a hole in a bound state.

flux: (magnetic) Lines of force used to represent magnetic induction.

fluxoids: Magnetic field lines that penetrate a superconductor, concentrate in regions, and interfere with the flow of current.

gauss: A unit of magnetic flux density.

homopolar generator: A low-voltage DC generator based on Faraday's disk device. The poles presented to the armature are all of the same polarity so that the voltage generated has uniform polarity producing a pure direct current.

intermetallic: An alloy of two metals that, when combined, form a different crystalline structure from that of the original metals.

Josephson effect: When electron pairs pass through a thin, insulating barrier between two superconductors without resistance.

Josephson junction: A superfast, superconducting electronic switch based on the Josephson effect.

lattice: The regular, periodic arrangement of atoms, ions, or molecules that form crystals.

maglev system: Magnetically levitated transportation, as in a train, that relies on the attractive or repulsive properties of magnets to support and/or propel the vehicle.

magnetic resonance imaging (MRI): A medical diagnostic system (based on the phenomenon of nuclear magnetic resonance) that uses protons to form an image of human body tissues.

magnetohydrodynamics: The generation of electric power by forcing ionized combustion gases through a superconducting magnet.

magtrain: A magnetically levitated train.

Meissner effect: The expulsion of lines of magnetic force from a material as it undergoes the transition to the superconducting phase.

Meissner motor: An electric motor developed at Argonne National Laboratory, its operating principle is based on the Meissner effect and it uses superconducting ceramic disks.

Meissner test: A small magnet is placed over a ceramic disk in liquid nitrogen; if the magnet floats, that is evidence of superconductivity.

metalloid: An element with both metallic and nonmetallic properties.

microbridge: In a superconducting film, a constriction that controls a device's applications.

perovskite: A group of natural and synthetic metallic oxides which have a three-dimensional crystalline structure.

phonon: A unit of vibrational energy in a crystal lattice.

photon: Particles of light.

planes: Two-dimensional links of copper and oxygen atoms that form the ceramic lattice.

plasmon: A subatomic particle.

quadrupoles: In an accelerator, magnets that focus the beam to a narrow diameter.

resonant frequency: The tendency of protons to reemit radio signals which have been carefully pulsed to match the protons' natural frequency.

sintering: Compacting a powdered material while heating it to a temperature below its melting point in order to weld the fine particles together into coarse lumps.

solenoid: A current-carrying coil of insulated wire which produces a magnetic field within the coil.

SQUID: For Superconducting Quantum Interference Device, used to measure tiny magnetic fields.

tesla: A unit of magnetic flux density; equal to 10,000 gauss.

TeV: A trillion electron volts.

tokamak: A toroidal magnetic "bottle" used to contain plasma for nuclear fusion.

ENDNOTES

CHAPTER 1: GOING FOR THE COLD

Page 1. **"I'm not wasting any more time . . ."**: "Superconductivity: Fact vs Fiction," by Karen Fitzgerald, in *IEEE Spectrum*, May 1988.

Page 4. **Sir Arthur Stanley Eddington**: *From Stars and Atoms*, Yale University Press, quoted in *A Treasury of Science*, Harper & Brothers, New York, 1958. 44.

CHAPTER 2: THE NATURE OF CURRENTS

Page 25. **Daniel Prober**: "Mystery of Electron Pairs," Yale University news release, July 7, 1987.

CHAPTER 3: THE QUEST HEATS UP

Page 33. **"Like other men . . ."**: *Familiar Medical Quotations*, Edited by M. B. Strauss. Little, Brown, 1968. 532.

Page 33. **Donald U. Gubser**: Personal interview.

Page 34. **Bernd Matthias**: "High Temperature Superconductivity?" In *Comments on Solid State Physics* 3, (no. 4, Oct/Nov 1970): 95.

CHAPTER 4: THE MATERIALS REVOLUTION

Page 35. **New Scientist**: "Magic Carpets and Electric Skis." Stanford University news release, July 7, 1987.

Page 36. **"I find it very, very satisfying . . ."**: ibid.

Page 36. **"They began lining up . . ."**: "Superconductors!" in *Time*, May 11, 1987. 62.

Page 38. **meeting of the Materials Research Society**: Personal interviews.

Page 42. **As a matter of fact . . .**: "Superconductivity at Room Temperature," by W. A. Little, in *Scientific American*, February 1965. 21.

Page 43. **I was struck**: Ibid. 22–23.

Page 44. **A graduate student in Hideki Shirakawa's . . .**: "Plastics That Conduct Electricity," R. B. Kaner, Alan G. MacDiarmid in *Scientific American*, February 1987. 106.

Page 49. **"We didn't tell anybody what we were doing . . ."**: "Superconductivity," by Gina Maranto in *Discover*, August 1987. 26.

Page 52. **"Our group read the paper . . ."**: *Time*, May 11, 1987. 67.

Page 53. **Gordon Pike**: *Discover*, August 1987. 27.

Page 55. **"Third World Lab"**: "Championing Chu," by Deborah Fowler, in *Texas Business*, November 1987. 31.

Page 55. **"We are all friends here . . ."**: Ibid.

Page 56. **"We were so excited . . ."**: *Time*, May 11, 1987. 68.

Page 59. **Robert C. Dynes**: "Signs of a New High in Ceramic Superconductivity," *Science News*, December 5, 1987. 356.

Page 60. **"This is a scientific breakthrough . . ."**: "Superconductivity Found at a Temperature Much Higher than Water's Boiling Point." *Metal Working News*, December 7, 1987.

Page 61. **"I've got to keep my priorities in order."**: "His Was the Little Laboratory That Could," by David Stipp in the *Wall Street Journal*, March 31, 1988.

CHAPTER 5: GETTING DOWN TO THE WIRE

Page 62. **"Thus . . . it is important for Japan . . ."**: *Science and Technology in Japan*, November/December 1987. 42.

Page 62. **Shoji Tanaka**: Personal interview.

Page 63. **"The race to commercialize . . ."**: Address by John S. Herrington before the Federal Conference on Commercial Applications of Superconductivity, Washington, D. C., July 29, 1987.

Page 63. **"When the transistor was invented . . ."**: Personal interview.

Page 64. **Masaki Suenaga**: "Experience Counts in Applying New Materials," *Brookhaven Bulletin*, December 11, 1987.

Page 66. **"We've improved the material's strength . . ."**: Personal interview.

Page 67. **Another novel approach**: "Making the New Superconductors Machinable." Ohio State University press release, March 23, 1988.

Page 68. **"The English said . . ."**: "Add Strontium to the Warm Recipe," *Financial Times*, August 28, 1987. 12.

Page 68. **"You start with any cheap oxide . . ."**: "Explosively-formed Superconductors Being Developed," by Sam Jones in *Metalworking News*, September 28, 1987.

Page 69. **"It's a very basic process . . ."**: "Superconducting Films Made Easy with Lasers." Bellcore news release, June 4, 1987.

Page 70. **Certainly, there is no lack of enthusiasm . . .**: Testimony of Dr. Praveen Chaudhari before the U.S. House of Representatives Science, Space and Technology Committee, June 10, 1987.

CHAPTER 6: CURRENT PROBLEMS

Page 72. **"All kinds of switches . . ."**: Personal interview.

Page 79. **"It doesn't take all that long . . ."**: Personal interview.

Page 82. **"Our single crystals had current densities . . ."**: Personal interview.

Page 83. **"You've just got to distinguish . . ."**: Personal interview.

CHAPTER 7: CHAINS, PLANES, AND GRAINS

Page 85. **Don Gubser**: Personal interview.

Page 89. **"I read an article somewhere . . ."**: Personal interview.

Page 90. **Benjamin Franklin and Mme. Marie Curie . . .**: *Physics*, by the Physical Science Study Committee, D. C. Heath & Company, 1960. 5.

Page 97. **Over the next few months**: In *Logos*, Argonne National Laboratories, 5 (no. 3, Autumn 1987).

CHAPTER 8: ON A MORE PRACTICAL NOTE

Page 101. **Goma Shobo**: "Laughing Matters: The Japanese Race Ahead in Superconductivity Lore." *Scientific American*, July 1988. 114.

Page 103. **You don't even have to make cars . . .**: "Levitation: Science Brings Magic to Life," by James Gleick in *The New York Times*, July 7, 1987, Science Times section.

Page 103. **"We do feel . . ."**: Personal interview.

Page 104. **"Maybe we're driving the field too hard . . ."**: Personal interview.

Page 104. **"If you want to talk about an agency . . ."**: Personal interview.

CHAPTER 9: SHIPPING OUT WITH SUPERCONDUCTIVITY

Page 115. **Yoshiro Saji**: Personal interview.

Page 118. **Stewart Way**: Personal interview.

Page 122. **"On the surface . . ."**: Personal interview.

Page 123. **Mike Superczynski**: Personal interview.

CHAPTER 10: FLYING TRAINS

Page 133. **"Both the air and highway modes . . ."**: Presentation to the Department of Energy on the application of high-temperature superconductivity to magnetically levitated trains by Larry Johnson, July 13, 1987.

Page 134. **At speeds of about 120 miles per hour . . .**: Ibid.

Page 138. **"It's not a locomotive technology . . ."**: Personal interview.

Page 142. **"I think the reason . . ."**: From an interview with Masanori Ozeki in *New Technology Week*, November 30, 1987.

Page 142. **The Japanese government . . .**: From a letter to constituents from Sen. Daniel Patrick Moynihan, April 10, 1988.

Page 144. **A casual observer can see . . .**: Testimony of Gordon Danby before the Senate Environment and Public Works Committee, February 26, 1988.

Page 145. **"If all of U.S. transport . . ."**: Testimony of James Powell before same committee.

CHAPTER 11: IS THERE A SUPERCONDUCTING CAR IN YOUR FUTURE?

Page 150. **Robert Eaton**: General Motors news release, February 2, 1988.

CHAPTER 12: SUPERCONDUCTING TECHNOLOGY

Page 156. **"We hope to start taking orders . . ."**: "Whatever Happened to Superconducting Generators?" *IEEE Spectrum*, January 1987. 17.

Page 159. **Results to date . . .**: "EPRI Cancels its Superconducting Generator Funds," in *Electric Light and Power*, October 1, 1983.

Page 160. **John Hulm**: Personal interview.

Page 167. **MHD is currently in the doghouse . . .**: Personal interview.

Page 173. **In the last analysis . . .**: Personal interview.

CHAPTER 13: MAGNETIC FIELDS AND THE BODY

Page 179. **"If people with the most problems . . ."**: Personal interview.

Page 182. **Samuel Williamson**: Personal interview.

Page 186. **Clearly, we need . . .**: Personal interview.

CHAPTER 14: THE SUPERCONDUCTING SUPERCOLLIDER

Page 189. **The cyclotron started a revolution . . .**: Speech by John Herrington before the National Symposium on the SSC, Denver, December 3, 1987.

Page 190. **"If automotive technology had progressed . . ."**: Remarks by President Reagan in a meeting with supporters of the SSC, March 30, 1988.

Page 191. **Lawrence Berkeley Laboratory**: "Supermagnets," in the *Lawrence Laboratory Research Review*, 10 (no. 2, Summer 1985).

Page 192. **"It's a bit like finding out . . ."**: "Worlds Within the Atom," by John Boslough, in *National Geographic Magazine*, May 1985.

Page 195. **"It took nearly 20 years . . ."**: "Will High Temperature Superconductivity Affect the SSC's Design?" by Irwin Goodwin in *Physics Today*, August 1987. 51.

Page 195. **Philip Anderson**: "Taking A Quantum Leap," by David Smith in *Bostonia Magazine*, July/August 1988. 48.

Page 196. **"A year ago . . ."**: "Will High Temperature Superconductivity Stop the SSC?" Letter from J. R. Fanchi in *Physics Today*, November 1987. 13.

Page 196. **As I reflect back . . .**: "What Price Funding the SSC?" Letter from J. F. Waymouth in *Physics Today*, July 1988. 10–11.

Page 197. **Joseph F. Salgado**: Testimony before the House Space and Technology Committee, March 27, 1988.

CHAPTER 15: JAPAN VS. AMERICA

Page 201. **Congressional Office of Technology Assessment**: Report, U.S. Government Printing Office, June 1988.

Page 205. **"The Japanese production advantage . . ."**: "How to Beat Japan at Its Own Game," by M. Borrus in *The New York Times*, July 31, 1988. F3.

Page 205. **Mohammad Sadli**: In *Asia Quarterly*, March 1982.

Page 209. **"The Japanese regard themselves . . ."**: "Japan Sets Its Sights For Research," by C. Clive Bradley in *The New Scientist*, September 29, 1988. 56.

Page 212. **"You just can't say . . ."**: "Levitation," by James Gleick in *The New York Times*, July 7, 1987.

Page 212. **"It is vital to America's international competitiveness . . ."**: News release, Argonne National Laboratory, October 5, 1988.

Page 213. **"There's no doubt these materials . . ."**: "MIT Moves to Commercialize New Superconductors," by Alan Cooperman for The Associated Press, November 30, 1988.

INDEX